Math For
Real Life

FOR

DUMMIES®

Math For Real Life
FOR DUMMIES

by Barry Schoenborn

WILEY

John Wiley & Sons, Inc.

Math For Real Life For Dummies®

Published by
John Wiley & Sons, Inc.
111 River St.
Hoboken, NJ 07030-5774
www.wiley.com

Copyright © 2013 by John Wiley & Sons, Inc., Hoboken, New Jersey

Published by John Wiley & Sons, Inc., Hoboken, New Jersey

Published simultaneously in Canada

For general information on our other products and services, please contact our Customer Care Department within the U.S. at 877-762-2974, outside the U.S. at 317-572-3993, or fax 317-572-4002.

For technical support, please visit http://www.wiley.com/techsupport.

Wiley publishes in a variety of print and electronic formats and by print-on-demand. Some material included with standard print versions of this book may not be included in e-books or in print-on-demand. If this book refers to media such as a CD or DVD that is not included in the version you purchased, you may download this material at http://booksupport.wiley.com. For more information about Wiley products, visit www.wiley.com.

Library of Congress Control Number: 2012955837

ISBN 978-1-118-45330-8 (pbk); ISBN 978-1-118-45332-2 (ebk); ISBN 978-1-118-45334-6 (ebk); ISBN 978-1-118-45331-5 (ebk)

Manufactured in the United States of America

SKY10057527_101323

About the Author

Barry Schoenborn lives in Nevada City, California. He's a longtime math, science, and technical writer, with over 35 years' experience. He's written hundreds of user manuals. In the past, Barry's technical writing company worked with the State of California agency CalRecycle to teach scientists and administrators how to write clearly.

Barry's the coauthor of *Technical Math For Dummies, Medical Dosage Calculations For Dummies, Physician Assistant Exam For Dummies,* and *Storage Area Networks: Designing and Implementing a Mass Storage System.*

He was a movie reviewer for the Los Angeles *Herald-Dispatch* newspaper and wrote a monthly political newspaper column for *The Union* newspaper of Grass Valley, California, for seven years. He also writes "dental humor," a funny genre that nobody except a dentist would want to read.

Barry's publishing company, Willow Valley Press, published *Dandelion Through the Crack,* which won the William Saroyan International Prize for Writing.

Dedication

I dedicate this book to Lynda Straus, my long-time intimate friend. She is one of the brightest people I know and is an excellent technical writer. However, she stays very busy, mostly acting as caregiver for her mother and has never made math a focus. Oops! In many ways, she is my model for the ideal reader of this book.

Author's Acknowledgments

I want to thank Lindsay Lefevere, Executive Editor, for the opportunity to write this book. A big thanks, too, to Matt Wagner of Fresh Books Literary Agency, who presented me to Wiley.

Many thanks to the Wiley team: Tracy Barr, Project Editor, and Michael McAsey and Shira Fass, the technical reviewers. They worked very hard to make this book more accurate and easier to read. Without them, there wouldn't be a book.

Publisher's Acknowledgments

We're proud of this book; please send us your comments at http://dummies.custhelp.com. For other comments, please contact our Customer Care Department within the U.S. at 877-762-2974, outside the U.S. at 317-572-3993, or fax 317-572-4002.

Some of the people who helped bring this book to market include the following:

Acquisitions, Editorial, and Vertical Websites

Editor: Tracy L. Barr

Executive Editor:
Lindsay Sandman Lefevere

Assistant Editor: David Lutton

Editorial Program Coordinator: Joe Niesen

Technical Editors: Michael McAsey,
Shira Fass

Senior Editorial Manager: Jennifer Ehrlich

Editorial Manager: Carmen Krikorian

Editorial Assistant: Alexa Koschier

Cover Photos: © PixelEmbargo /
iStockphoto.com

Cartoons: Rich Tennant
(www.the5thwave.com)

Composition Services

Project Coordinator: Patrick Redmond

Layout and Graphics: Carrie A. Cesavice,
Joyce Haughey, Melissa Smith,
Christin Swinford

Proofreader: Wordsmith Editorial

Indexer: BIM Indexing & Proofreading
Services

Publishing and Editorial for Consumer Dummies

    Kathleen Nebenhaus, Vice President and Executive Publisher

    David Palmer, Associate Publisher

    Kristin Ferguson-Wagstaffe, Product Development Director

Publishing for Technology Dummies

    Andy Cummings, Vice President and Publisher

Composition Services

    Debbie Stailey, Director of Composition Services

Contents at a Glance

Table of Contents

Introduction

● ●

Math is great. I've spent my life enjoying math and getting the benefits from it. But many people don't think math is so hot: They have fear and loathing.

The trouble is, a lot of what you learned in school probably was boring or painful, and most likely you forgot it right away. Or you may have deliberately *not* learned some types of math at all just to save yourself the trouble of forgetting them later. Why? Mainly because a lot of the math taught in schools wasn't math you could use. Well, that nonsense stops here, because this book is filled with math you *can* use. Here's just a brief sampling of the areas where you'll find real-life math to be useful:

- ✔ At home, math comes in handy in the house, yard, workshop, and hobby room. It's also a big part of cooking in the kitchen, whether you're following a recipe or counting sticks of celery.

- ✔ The grocery store and shopping center are the places where most people buy the most items with the most frequency. You can use math when you're shopping to make better choices and get better deals.

- ✔ Real-life math helps with understanding food labels, losing weight, and exercising. You can get healthy by chance, but choice is better, and math helps you make choices.

- ✔ On the job, you'll find that a brush-up on math skills is very handy, whether you're filling out a timesheet, managing time on a project, making change, or doing any other math-related tasks.

About This Book

Math for real life is math you need, because math for real life is math you *use*. And that's what I focus on in this book, which is different from other math books. Here's a quick rundown of the major differences:

- ✔ This book is all about *practical* math. Although I cover math fundamentals (which are the building blocks of math for real life), I quickly move to problems you might deal with every day and the specific math skills you need to handle them. Other math books are often filled with abstractions.

✔ The book doesn't include any high-level equations, because you don't need them to go shopping or to calculate your auto repair bill. Even the world of gambling operates on simple formulas.

✔ It takes a *comprehensive* look at applying math in real-life areas. I include a little bit about a lot of subjects, and no subject goes any deeper than you need it to go. Many books are devoted to one subject (for example, algebra, geometry, trigonometry, or pre-calculus). Not this one — it's devoted to everything.

✔ It's *not dull* (I hope), as other math books often are. Because it's a *For Dummies* book, you can be sure that it's easy to read and has touches of humor.

But wait! There's more! At the risk of sounding like a late-night infomercial, you'll find other unique features that I gar-on-tee you won't find in a more traditional math book:

✔ You get terms, definitions, and word origins. The reason is that topics such as investments and insurance use so many odd words — and they need to be defined and explained.

✔ You get special insights into our culture and the way we think. For example, a "sale" isn't always a sale, and "free" doesn't always mean free. Even so, we tend to buy. Math for real life shows you where the exaggerations are.

Conventions Used in This Book

This book is user-friendly: easy to hold in your hands, easy to read, and easy to understand. On top of that, it's easy to navigate, too, because the table of contents, the index, and the "In This Chapter" section at the beginning of each chapter help you find information you're looking for.

The book uses the following conventions:

✔ *Italic type* highlights new terms. Once in a great while, you'll see italics used for emphasis.

✔ Although English teachers would cringe at my breaking the rules, I usually write numbers as numerals, not words. For example, the text will say "if you drive 30 miles on 2 gallons of gas," not "if you drive thirty miles on two gallons of gas."

✔ *Variables* in formulas appear as italics (for example, $3a + 4b = 10$).

✔ Web addresses are in monofont. They are usually very short and shouldn't break across two lines of text. But if they do, no extra characters indicate the break. Just type what you see into your browser.

What You're Not to Read

It would be great if you read all the words of this book in the order they appear, but life is short. You don't have to read chapters that don't interest you. This is a reference book, and it's designed to let you read only the parts you need. And if you get stuck, then you can go over to a chapter you skipped to get some help.

Here's a short list of "skippable" information. Information in these bits isn't essential to doing real-life math:

✔ You don't have to read anything with a Technical Stuff icon next to it. That text gives you a little extra information about a technique, the origin of a principle, or maybe a formal definition.

✔ Sidebars (that's what they're called in publishing) are blocks of text with a gray background. They are interesting (I think), but aren't critical to your understanding the main text.

Foolish Assumptions

The book makes some assumptions about you and what you're looking for in a math book:

✔ **You were exposed to math fundamentals in *elementary school* but may have forgotten a few of them.** (Why not high school? Because in high school many people get bored, dazed, or frustrated with mathematics. So although you may have been in class, your mind was probably somewhere else.) Even if you missed some basic math concepts in school, don't fret: I review most of them in this book.

✔ **You're only interested in information that's relevant to you and are likely to skip concepts you're already comfortable with.** That's okay. This is a reference book, not a novel.

✔ **You have access to a computer and the Internet.** Although not essential, being able to access the Internet is very handy. You can use a search engine to find useful specialized calculators or to learn more about any topic in this book.

How This Book Is Organized

This book has four parts, each representing a particular math topic.
The chapters in the parts focus on different aspects of that topic.
Overall, the book moves from an early review of basics to chapters
about math that comes up in everyday life to topics related to personal finance. Of course, you don't have to read the chapters in the
order they appear. Following is an overview of the kind of information you can find in each part.

Part 1: Boning Up on Math Basics

In this part, you get math basics, which amount mostly to counting and simple arithmetic. Chapters 1, 2, and 3 bring out broad
concepts related to the arithmetic fundamentals. In Chapter 4, you
work with simple and useful statistics. Chapter 5 is about mental
math, a great shortcut when you don't have a calculator handy
(which is most of the time).

Part II: Math for Everyday Activities

Part II shows you how to do the calculations that spring up regularly
in real life. Want to seed your lawn or plant a flower bed? Math
is involved. How about preparing a dinner for six from a recipe
designed to feed four? You need math for that, too. Ever tried to
decide whether the higher-priced but bigger box of cereal is a
better deal than the lower-priced but smaller box? Again, math
comes to the rescue. Shopping, cooking, driving around town,
dining out, or trying to lose weight — math makes all these tasks
easier.

Part III: Math to Manage Your Personal Finances

True, you may deal with personal finances daily, but they represent
a sort of "special" kind of math. To handle these tasks well, you
need to understand some general principles, a few specialized
terms, and a few strategies. Fortunately, this part has you covered.
Here you can get info to create a budget, better manage your bank
account and check register, avoid credit card debt, invest more
wisely, and more.

Part IV: The Part of Tens

What better way to end a book chock-full of easy-to-apply math formulas and principles than a couple of lists highlighting calculations you can do in your head and games you can play to build your math skills and sharpen your critical thinking? Consider this the icing-on-top-of-the-cake part.

Icons Used in This Book

In the margins of this book you'll see small drawings called *icons*. Each icon calls out a special kind of information.

 A tip is a suggestion or a recommendation. It usually points out a quick and easy way to get things done or provides a handy piece of extra information.

 A warning alerts you to conditions that, if you're not careful, could lead you to wrong answers, faulty conclusions, or otherwise mess up your day.

 This icon appears beside information that's important enough to keep in mind, both for the task at hand and in general.

 I use this icon to share esoteric or otherwise interesting but non-essential information.

Where to Go from Here

You can go to any chapter of the book from here. Although I've written this book so that the basic info comes first, you can start anywhere you want. Need a little more guidance? Here are some suggestions:

- ✔ If you're browsing for a topic that piques your interest, check out the table of contents. Here you can see all the topics this book covers. Chances are one (or more) will call to you immediately. You can also try the "thumb test": Riffle through the pages until something catches your eye.

- ✔ If you haven't made a choice, begin with Chapter 1. It has broad concepts and is a good launching pad into the discussions elsewhere in the book.

✔ If you have a particular problem (for example, maybe with shopping or investments), head to the table of contents or the index to find what you're looking for.

If you get stuck at any time, you'll probably find another chapter that can help you out. Just stop what you're reading and go visit that chapter.

Part I

Boning Up on Math Basics

In this part . . .

*I*n this part, you'll find a review of math basics, including simple math concepts from your school days, like numbers, counting, and arithmetic operations. You also discover math principles like ratio-proportion, conversions, and statistics and probability. I also share the best all-purpose calculation method. Chapter 5, which is all about doing simple math in your head, is a math bonus. Mental math is a handy tool, and the techniques aren't hard to learn.

Chapter 1

Awesome Operations: Math Fundamentals

In This Chapter

▶ Reviewing the four arithmetic operations

▶ Manipulating fractions

▶ Using charts to convey and understand information

▶ Strategies to help you solve word problems

*M*ath has basic operations that you need to know. These operations — addition, subtraction, multiplication, and division — make all the other math in this book possible.

The good news is that you most likely learned about basics (like counting) even before you entered school, and you learned about basic arithmetic operations in elementary school. So you've been at it for a long time.

In this chapter, I review counting and the fundamentals of the four basic arithmetic operations. Other important topics I cover here are fractions, percentages, charts and graphs, and word problems. But don't worry: None of these are mysterious.

Numbers You Can Count On

The most fundamental component of math is numbers. The first thing you do with numbers is count, and you probably started counting when you were very young. As soon as you could talk, your mother cajoled you to tell Aunt Lucy how old you were or to count from 1 to 5.

Counting was the first and most useful thing you did with math, and you still use it every day, whether you're buying oranges at the grocery store or checking the number of quarts of motor oil in a case.

Counting has been essential since people first walked the earth. In fact, the Ishango bone is a tally stick (a counting stick), and it's over 20,000 years old!

Several kinds of numbers exist. Over time, mathematicians have given them many names. The two most important kinds are whole numbers and fractions. To see a little bit about how these numbers work, use a *number line*, a simple display of numbers on a line (see Figure 1-1).

Illustration by Wiley, Composition Services Graphics

Figure 1-1: A number line.

The numbers to the right of 0 are called *natural* numbers or *counting* numbers. Of course, they are the numbers you use to count. They're easy for anyone to work with because they represent how many of something someone has (for example, 6 apples or 3 oranges).

Over many centuries and in different cultures, people made up the number 0, which represents the lack of a quantity. The numbers to the left of 0 on the number line, *negative* numbers, are a harder concept to grasp. You recognize negative number in real life. For example, if your checking account is overdrawn, you have a negative balance. If someone owes you $3.00, you have "negative cash" in your pocket.

Here are the key points to know about the number line:

- ✔ All the numbers you see in Figure 1-1 are *whole* numbers, also called *integers*. An *integer* is a number with no fraction part. The word comes from Latin, and it means "untouched," so it's the whole deal.

- ✔ The numbers to the right of zero are *positive integers*. The numbers to the left of zero are *negative integers*.

 Mathematicians (and I'm not making this up) have trouble with zero. The best they can do is attach it to the positive integers and label the group *non-negative integers*.

- ✔ The number line stretches to the left and right, to infinity and beyond (as Buzz Lightyear says).

- ✔ Decimals (such as 0.75) and regular fractions (such as 3/5) are only a part of a whole number. They all have a place somewhere on the number line. They fit in between the integers. For example 2.75 "fits" between 2 and 3 on the number line, because it's greater than 2 but less than 3.

Reviewing the Four Basic Operations

To do any sort of math, you need to know your math basics. The four basic operations — addition, subtraction, multiplication, and division — let you take care of all kinds of real life math. But what's also very important is that those same basic math operations allow you to handle fractions and percentages, which come up all the time in ordinary math tasks. Later (in Chapter 2), these operations form the basis for managing algebra equations and geometry.

The core operations are addition and subtraction. You very likely know what they are and how they work. Multiplication and division are "one step up" from addition and subtraction. The following sections give you a quick review of these four operations.

Addition

Addition is a math operation in which you combine two or more quantities to get (usually) a larger quantity. Addition was probably the first math you ever did.

You can add numbers (called the *operands*) in any order. This property (that is, the ability to perform the operation in any order) is called *commutativity*.

$$21+31+41+51=144$$
is equal to
$$51+41+31+21=144$$

No matter in what order you add the operands, the sum still equals 144.

Subtraction

Subtraction is a math operation in which you take away the value of one number from another, resulting in (usually) a smaller quantity.

In subtraction, the order of the operands is important. You can't rearrange the numbers and get the same answer. For example, $77 - 22$ (which equals 55) is not the same as $22 - 77$ (which equals -55).

Multiplication

Think of multiplication as repeated addition. For example, you likely know that $3 \times 4 = 12$, but you can also get there by adding 3 four times:

$3 + 3 + 3 + 3 = 12$

The technique also works for large numbers. For example, $123 \times 7 = 738$ is equivalent to this:

$123 + 123 + 123 + 123 + 123 + 123 = 738$

But who wants to do all that adding?

Here's the best advice for multiplication:

✔ For little numbers, know your multiplication table. It's easy, up to 10×10.

✔ For big numbers, use a calculator.

As with addition, you can multiply the numbers in a list in any order. The expression 3×4 is the same as the expression 4×3. Both equal 12.

Division

Division is essentially "multiple subtraction." In a simple problem such as $12 \div 4 = 3$, you can get the result by subtracting 3 four times from the number 12.

$12 \div 3 = 4$ with no remainder

is equal to $12 - 3 - 3 - 3 - 3 = 0$ (4 subtractions with no remainder)

In division, the order of the operands is important. You can't rearrange them and end up with the same answer.

Finagling Fractions

Fractions take several forms, but in real life, the forms you deal with are common fractions and decimal fractions.

A *common fraction* has two parts. The *numerator* is the top number, and the *denominator* is the bottom number. You don't have to learn these words, however. Just think "top number" and "bottom number."

$$\frac{\text{numerator}}{\text{denominator}}$$

What do you do with fractions? Arithmetic operations and conversions, that's what.

A common fraction is sometimes called a *simple fraction* or a *vulgar fraction.* The vulgar fraction isn't really rude; *vulgar* is just another word for *common* (from the Latin *vulgus,* meaning "common people").

Getting familiar with types of fractions

Like the popular ice cream parlor, fractions come in several flavors. Not 31 flavors, however. For this book, you have to remember only a few fraction types:

- **Proper fraction:** In a *proper fraction,* the numerator is smaller than the denominator (for example, $\frac{4}{9}$).

- **Improper fraction:** In an *improper fraction,* the numerator is larger than the denominator (for example, $\frac{9}{4}$). Think "Honey, does this numerator make my fraction look big?"

- **Mixed fraction:** A *mixed fraction* is a combination of a whole number and a fraction. Here's an example of a mixed fraction:

  $1\frac{3}{4}$

- **Decimal fraction:** A *decimal fraction* uses a decimal point (for example, 0.23, 1.75, or $47.25).

Decimals are fractions, too, even though they don't look like the other types of fractions. Look at this: 0.75 is a decimal. But what does that really mean? It means 75/100.

Reducing fractions

Here's fair warning: Doing fraction math often produces "clumsy" fractions. By clumsy, I mean unwieldy proper fractions (48/60, for example) and bad-looking improper fractions (37/16, for example). They are handy during the calculations but are very inconvenient as final answers.

You turn a clumsy fraction into something lovely to behold by *reducing* it.

Reducing proper fractions

You reduce proper fractions by finding a number that the numerator and denominator share and then separating it out. This tactic is called *factoring*, and multiplication rules allow you to do it. For example, for the fraction $\frac{48}{60}$, you "break out" the common factor 12 in both the numerator and denominator:

$$\frac{48}{60} = \frac{4 \times 12}{5 \times 12}$$

$$\frac{48}{60} = \frac{4}{5} \times \frac{12}{12}$$

$$\frac{48}{60} = \frac{4}{5} \times 1$$

$$\frac{48}{60} = \frac{4}{5}$$

When a fraction has the same numerator and denominator, it's equal to 1. Hence, $\frac{12}{12}$ becomes 1.

Another way of describing this is to say, "You reduce a proper fraction by dividing the top and bottom numbers by the same number."

Reducing improper fractions

To reduce an improper fraction, you break it into whole numbers and a remaining, smaller fraction. To do this, you divide the top number by the bottom number, and then you use the whole number and the remaining fraction to form a mixed fraction. Here's an example:

$$\frac{49}{16} = \frac{16 + 16 + 16 + 1}{16}$$

$$\frac{49}{16} = \frac{16}{16} + \frac{16}{16} + \frac{16}{16} + \frac{1}{16}$$

$$\frac{49}{16} = 1 + 1 + 1 + \frac{1}{16}$$

$$\frac{49}{16} = 3\frac{1}{16}$$

Adding, subtracting, multiplying, and dividing fractions

Fractions are just numbers. Like integers, you can add, subtract, multiply, and divide them. Before you panic, keep in mind that you perform these math calculations on fractions all the time. Don't believe me? Think about money.

At first, dollars and cents don't look like fractions because they're in decimal form. But they are fractions, for sure. To look at the details, take a gander at the following sections.

Addition

To add two fractions, the fractions must have the same denominator (also called a *common* denominator). After the denominators are the same, you add fractions simply by adding the numerators.

When the denominators aren't the same, you need to make them the same. You can't directly add 1/2 pie to 1/4 pie to get 3/4 pie, for example. You need to convert the 1/2 pie into quarters (2/4 pie). Figure 1-2 shows what adding pieces of pie looks like.

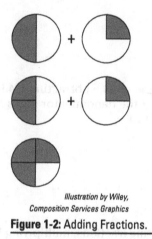

*Illustration by Wiley,
Composition Services Graphics*

Figure 1-2: Adding Fractions.

Getting the denominators the same is easy because you're allowed to multiply both the top number and the bottom number by the same number. In the pie example, you multiply both numerator and denominator of the fraction 1/2 by 2:

$$x = \frac{1}{2} \times \frac{2}{2}$$

$$x = \frac{1 \times 2}{2 \times 2}$$

$$x = \frac{2}{4}$$

After you have all operands in 1/4 pie units, adding 2/4 and 1/4 to get 3/4 is easy. (Remember that the denominator stays the same when you add the numerators.)

Subtraction

To subtract two fractions, the fractions must have a common denominator (just as they must in addition); then you simply perform the operation on the numerators.

If the denominators aren't the same, you need to make them the same before you can subtract. For example, you can't directly subtract 1/4 pie from 1 whole pie (which in fraction form is 1/1) to get 3/4 pie because the denominators are different. Again, you need to convert the whole pie into quarters, and you do that by multiplying the numerator and denominator by 4 to get 4/4 pie. Then you can do the subtraction:

$$x = \frac{1}{1} \times \frac{4}{4}$$

$$x = \frac{1 \times 4}{1 \times 4}$$

$$x = \frac{4}{4}$$

After all the operands are in 1/4 pie units, subtracting 1/4 from 4/4 to get 3/4 is easy. (Remember that the denominator stays the same when you subtract the numerators.)

$$x = \frac{4}{4} - \frac{1}{4}$$

$$x = \frac{3}{4}$$

Multiplication

Compared to adding and subtracting fractions, multiplying fractions is easy. Just multiply the numerators, multiply the denominators, and then reduce.

$$x = \frac{3}{5} \times \frac{6}{7}$$

$$x = \frac{3 \times 6}{5 \times 7}$$

$$x = \frac{18}{35}$$

The answer is 18/35. When possible, try to reduce the result. In this case, you can't reduce 18/35 at all.

Division

Here's the secret to dividing fractions: *Invert and multiply.* That is, flip the second fraction so that the numerator is on the bottom and the denominator is on the top, and then multiply as you would any other fraction.

Say you want to divide 1/4 by 2. (***Note:*** The fraction form of a whole number is that number over 1.) The answer is obviously 1/8. Not so obvious, you say? Here's how you get the answer:

$$x = \frac{1}{4} \div 2$$
$$x = \frac{1}{4} \div \frac{2}{1}$$
$$x = \frac{1}{4} \times \frac{1}{2}$$
$$x = \frac{1}{8}$$

You follow the same process when you want to divide a fraction by a fraction:

$$x = \frac{1}{4} \div \frac{1}{3}$$
$$x = \frac{1}{4} \times \frac{3}{1}$$
$$x = \frac{3}{4}$$

Notice that dividing by a fraction yields a *higher* result than dividing by a whole number.

You can't divide by 0. It's mathematically impossible. The old saying is, "Never divide by zero! It's a waste of time, and it annoys the zero."

Converting fractions

The handiest fraction conversions are turning common fractions into decimal fractions and turning decimal fractions into common fractions.

A fraction is a ratio, too

In math, a *ratio* is a relationship between two numbers. I mention this because ratios come up all the time.

The size of a wide-screen DVD image is called the *aspect ratio,* usually 16:9. That amounts to 16 inches of width for every 9 inches of height, and it doesn't really matter how big your TV screen is. The ratio is always the same.

If you have a gas-powered weed eater (also known as a *string trimmer* or *weed whacker*), you've probably bought 40:1 2-cycle engine oil for it. The 40:1 ratio means that you mix 40 parts gas to 1 part oil.

Turning a common fraction into a decimal fraction

To turn a common fraction into a decimal fraction, just divide the denominator into the numerator. A number like 4/5 easily turns into 0.80 when you divide 4 by 5.

Don't be surprised or alarmed if some division doesn't come out "even." For example, the decimal equivalent of 1/3 is 0.333333333 (and the 3s go on forever). If you see a sale item marked "33% off," it's been reduced by 33 percent or about 1/3. If the item is marked "20% off," it's been reduced by 20/100, or 1/5. (See the section "Processing Percentages" for the lowdown on how to work with percentages.)

Turning a decimal fraction into a common fraction

To turn a decimal fraction into a common fraction, just express the decimal as a fraction and reduce the fraction.

A decimal with one decimal place (0.6, for example) needs a fraction with 10 in the denominator. A decimal with two decimal places (0.25, for example) needs a fraction with 100 in the denominator, and so forth. Here are some examples:

$$0.6 = \frac{6}{10}$$

$$0.71 = \frac{71}{100}$$

$$0.303 = \frac{303}{1000}$$

Notice that the number of zeroes in the denominator is the same as the number of decimal places in the decimal fraction.

For example, say you want to convert 0.375 into a fraction. Here's how you'd go about it:

$$x = \frac{375}{1,000}$$

$$x = \frac{3 \times 125}{8 \times 125}$$

$$x = \frac{3}{8} \times \frac{125}{125}$$

$$x = \frac{3}{8} \times 1$$

$$x = \frac{3}{8}$$

In this example, when you "factor out" 125 from both the numerator and denominator, the result is the common fraction 3/8. See the section "Reducing proper fractions" for details on factoring.

Processing Percentages

A percentage is a fraction whose denominator never changes. It's always 100. A number like 33 percent, for example, refers to 33 parts in 100, or 33/100, or 0.33. You see percentages written as "33%" and "33 percent." No matter how it's written, it's just another way of saying "thirty-three parts in one hundred."

Percent and *per cent* means "per centum," which is from the Latin phrase meaning "by the hundred." So a percentage always refers to a number of parts out of 100.

Percentages are especially handy for comparing two quantities. For example, if one beer contains 5.5 percent alcohol and another contains 12 percent alcohol, you can be sure that the "high octane" beer has a lot more punch.

Percentages also let you compare values to an arbitrary standard. Nutrition labels are a good example They compare items in food, such as dietary fiber, cholesterol, or vitamins and minerals, to the Dietary Reference Intake (DRI) nutrition recommendations used by the United States and Canada.

A percentage is a *dimensionless proportionality,* meaning that it doesn't have a physical unit. Fifty percent of a length is still 50 percent, whether you're talking about feet or light years.

Converting a common fraction to a percentage

Sometimes you want to convert a fraction to a percentage. Say, for example, that you're fed up with your commute to work, because the drive requires 1 hour each way. You're at the job for 9 hours, so work consumes 11 hours of your day, 2 of those hours with you sitting in traffic. While stuck in bumper-to-bumper traffic, you wonder what percentage of your work-related time in spent commuting. The fraction is 2/11, so what's the percentage?

To convert a common fraction into a percentage, just divide the numerator by the denominator and multiply the result by 100:

$$\text{percentage} = \frac{2}{11} \times 100$$
$$\text{percentage} = 0.1818 \times 100$$
$$\text{percentage} = 18.18$$

You can see that 2/11 is about 18 percent. What could be simpler than that?

A percentage is a ratio, too

As I mention earlier, a ratio is a relationship between two numbers. A percentage can often be expressed as a ratio. For example, if a bottle of vodka contains 40 percent alcohol (which, confusingly, is called 80 proof in the United States), that means that 40 parts in 100 are alcohol. That's a ratio of 40:60, 40 parts of alcohol to 60 parts of water.

You can convert from a ratio to a percentage, too. For example, a "four to one" martini has a gin:vermouth ratio of 4:1. The vermouth is 1/5 of the cocktail, or 20 percent.

Converting a percentage to a fraction

Sometimes a fraction may be more convenient than a percentage. Perhaps you want to know what fraction of your salary goes to taxes. Or maybe you're less inclined to eat a whole 8-ounce bag of chips when you think in terms of it having 1/2 rather than 50 percent of your daily recommended amount of sodium.

To convert a percentage into a common fraction, just divide the percentage by 100 and reduce the result. For example, say to want to convert 80 percent into a common fraction:

$$80 \text{ percent} = ???$$

$$80 \text{ percent} = \frac{80}{100}$$

$$80 \text{ percent} = \frac{4 \times 20}{5 \times 20}$$

$$80 \text{ percent} = \frac{4 \times \cancel{20}}{5 \times \cancel{20}}$$

$$80 \text{ percent} = \frac{4}{5}$$

The value 80 percent means 80/100. Form a fraction and reduce it. As you can see, 80 percent is 4/5.

Grasping Charts and Graphs

A chart or graph is a *visual representation* of numbers. Charts and graphs come in many forms, but for day-to-day math, you need to know about only three kinds — the line chart (or graph), the pie chart, and the bar chart.

The key point is that a chart is visual, and people usually find a visual display to be more understandable than a list of numbers. Expect to encounter charts when you read about the economy or when you compare consumer products. Also, the best thing is that you can make your own charts, which you may want to do, for example, to get a better picture about your personal finances.

Looking at line charts

A line chart (sometimes called a *line graph*) displays information as data points connected by a line. With this chart, you can easily see how an item is trending. Figure 1-3 shows typical temperatures over a week. What can you glean from this data? That the weekend was hot!

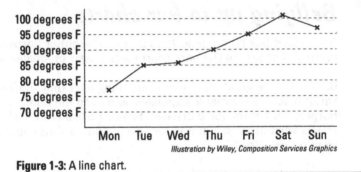

Illustration by Wiley, Composition Services Graphics

Figure 1-3: A line chart.

A line chart can easily show you how the economy is doing. Think unemployment figures. Also, you can make a chart that shows how one (or all) of your investments is doing.

Gobbling up pie charts

A pie chart looks like, er, a pie, which is divided into "slices" that show the relative proportion of various elements. This type of chart lets you see both the relationship between elements and the relationship of individual elements to the whole pie.

Pie charts are great when you have to compare only a few elements. When you must compare many elements, the slices get too thin and they're harder to understand.

Figure 1-4 shows a typical monthly budget. After paying the rent, making the car payment, and buying food, you can see that not much is left for everything else. Note that it doesn't matter whether you make $1,000 a month or $10,000 a month. The pie chart shows *relative* proportions.

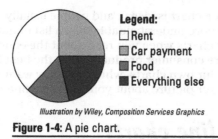

Legend:
☐ Rent
▨ Car payment
▤ Food
■ Everything else

Illustration by Wiley, Composition Services Graphics

Figure 1-4: A pie chart.

A pie chart is great for making comparisons of government expenses relative to each other. And seeing where your tax money goes is always fun. Visit the Center on Budget and Policy Priories: `http://www.cbpp.org/cms/index.cfm?fa=view&id=1258`.

Bellying up to bar charts

A bar chart has rectangular bars that can be either horizontal or vertical. The size of the bars represents bigger or smaller values.

Bar charts are great for showing anything over time, including variable income, variable expenses, and even the number of burgers sold at the local drive-in. Figure 1-5 is a bar chart that shows what my Visa bill was for seven months. Can you tell when I went on vacation?

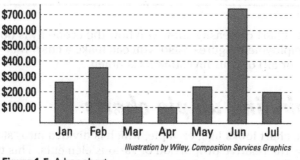

Illustration by Wiley, Composition Services Graphics

Figure 1-5: A bar chart.

Working Wicked Word Problems

Do you remember word problems (sometimes called "story problems") from school? A few people loved them, but many people hated them.

What's peculiar is that most of life's math problems start as word problems, a fact that's understandable because we speak in words, not numbers. So if you say, "The boss gave me a 10 percent raise," figuring out your new salary starts as a word problem.

At first glance, some word problems appear to be baffling. But that's just at first glance. You simply need to know a few tricks that can make all word problems easy to solve. The basic process for solving word problems is to first do some analysis and then do the math.

Doing the analysis

Two parts are involved in solving a story problem. The first part is to study the problem a little. That makes the second part (doing the math) easy.

For example, a shed has a roof that's 6 feet by 10 feet on each side. The barn's roof is twice as long and twice as deep as the shed's roof. Both buildings are red. If it takes 120 shingles to cover the shed's roof, how many shingles does it take to cover the barn's roof?

When you analyze a story problem, you go through the problem to get the info you need to eventually solve it. Follow these steps:

1. **Read the problem and list the facts.**

   Always read word problems more than once. Facts are hiding in the question. From the question, you know the dimensions of the shed roof *on each side*. You get a sense of the dimensions of the barn's roof, and you know how many shingles are needed to cover the shed. Good!

2. **Figure out exactly what the problem is asking for.**

   In every word problem, you run the risk of solving — correctly — for the *wrong* thing. So make sure you know what the question asks for. In the example, you know that the answer is "number of shingles to cover the barn." The question could have been about calculating the number of shingles to cover both the shed and the barn, but it's not.

3. **Eliminate excess information.**

   Both real life and school story problems tend to have extraneous facts. Ignore them. For example, the fact that both buildings are red is interesting but not important.

4. **See what information is missing.**

Sometimes a major fact is missing. What's more likely, however, is that the information is *hiding.* For example, the info that the barn's roof is twice as long and twice as deep as the shed's gives you a clue about calculating the area of the barn's roof.

5. **Find the keywords.**

Be on the lookout for key words and phrases, such as "how much more," "how much less," and "total." Those words and phrases usually indicate what kind of math operations are involved.

Applying the math

Almost every story problem uses a simple algebra formula that's "hiding" in it. When you develop the formula, you then insert the numbers to solve the problem. Math instructors often call this last step "plug and chug."

To apply the math, take the info you gleaned from your analysis and do the following:

1. **Convert information supplied into information needed.**

First, use the given dimensions of the shed roof to calculate how many square feet are covered by 120 shingles.

area (shed) = length × depth × 2

area (shed) = 6 × 10 × 2

area (shed) = 120

The answer is 120 square feet. (*Note:* You multiply by 2 to take into account *both* sides of the shed's roof.)

Then use the given dimensions of the shed roof to calculate the area of the barn's roof. The barn's roof is twice as long and twice as deep as the shed's roof.

area (barn) = 2(shed length) × 2(shed depth) × 2

area (barn) = 2(6) × 2(10) × 2

area (barn) = 12 × 20 × 2

area (barn) = 480

The answer is 480 square feet.

2. **Apply a formula.**

There's a technique called *ratio-proportion*. Don't worry about the details now (I explain it fully in Chapter 3). Here, you apply the technique:

$$\frac{\text{known quantity (shed area)}}{\text{known quantity (shed shingles)}} = \frac{\text{known quantity (barn area)}}{\text{desired quantity (barn shingles)}}$$

$$\frac{120}{120} = \frac{480}{x}$$

$$120x = 57{,}600$$

$$x = 480$$

You cross multiply and solve. The answer is 480 shingles.

Pay attention to units and phrase the answer in the units asked for. In the example, you must express the answer in shingles, not square feet.

3. **Check for reasonableness.**

   Always make sure the answer is reasonable. Because the barn is bigger than the shed, the barn should take more shingles than the shed. The 120 versus 480 is one reasonableness check. In the example, if you get an answer of 48 shingles or 48,000 shingles, something is wrong.

If you crave a shortcut, consider this: The fact that the barn's roof is twice as long and twice as deep as the shed's roof means that the barn's roof has four times the area. With that info at hand, the calculation is easy: Simply multiply the shed's 120 shingles by 4, giving you 480 shingles.

Other story problem tricks

If you find yourself totally stuck on a word problem, a few tricks may help you out:

- ✓ **Draw a diagram.** Sometimes, drawing a picture using the facts in the problem can be a help. This tactic works when you need to find the area of a garden, the board feet you need for a deck, or how old your brothers and sisters will be when you reach a certain age.

- ✓ **Find a formula.** When you encounter a problem about interest on your savings account or the amount of mortgage payments, chances are excellent that someone has already developed a formula to solve it. Chances are also very good that you can find an online calculator or an embedded function in a spreadsheet application to help you out.

✔ **Develop a formula.** Sometimes you can make your own simple formula instantly. For example, as soon as you know that a hamburger has 21 grams of protein and that dietary guidelines recommend 56 grams of protein a day, a little quick math (divide 56 by 21) shows that about three burgers at that backyard barbecue will give you a full day's worth of protein. And a formula works all the time, after you develop it.

✔ **Consult a reference.** Using a reference isn't just desirable; it's also sometimes necessary. For example, if you're painting a room, calculating the area to be covered isn't hard (see Chapter 8), but it's essential to consult the paint manufacturer's information to learn how much area a gallon of paint will cover.

Chapter 2

High School Reunion: Revisiting Key Principles of Algebra and Geometry

In This Chapter

▶ Understanding variables, constants, expressions, and equations

▶ Performing operations on algebraic equations

▶ Getting (re)acquainted with basic geometric shapes

▶ Using common formulas to determine area, perimeter, and volume

Did you and math part ways? If so, you likely stopped dating each other and broke up in high school, which is when most students meet up with algebra and geometry for the first time. You may have avoided these math classes, or maybe you took them but didn't pay as much attention as you should have.

As it turns out, algebra and geometry have some super concepts, and those concepts have enormous practical value, as you see in the rest of this book.

In this chapter, I take you back to high school for a reunion of sorts: I reintroduce you to some of the basic concepts and vocabulary of these two handy branches of mathematics. With those in hand, you'll be able to solve most algebra or geometry problems that come up in real life.

"A" Stands for "Algebra" and "Awesome"

Algebra is a branch of math that deals with variables and constants, and their relationship to each other in equations. When you solve a real-life math problem, chances are you're using algebra.

In this section, you get friendly with the names of the parts of an algebra statement. Then you go on to do simple (but essential) math operations. As you read through the next sections, keep these key points in mind; they may allay any trepidation you have about algebra:

✓ **In algebra, letters represent numbers — that's it.** And they represent numbers only until you solve the problem and replace them with numbers.

✓ **The operations used in algebra problems aren't mysterious.** You work on algebraic variables and constants, using the same math operations you use on numbers in "plain" arithmetic.

Getting acquainted with variables and constants

This sounds amazingly obvious, so get ready: Variables vary in value (until you determine what they are), and constants are constant in value.

Oh, you're so variable

A *variable* in algebra is a number whose value you don't yet know, so it's represented by a letter. The value of the variable may be anything, and that's why it's called a *variable*. The following are typical variables:

$$a \quad b \quad c \quad x$$

In an algebraic statement, you often see more than one variable and maybe a number or two. Here's an example:

$$a + 2 = b$$

You read and say the statement in almost the same way you'd say, "1 plus 2 equals 3." Instead you just say "a plus 2 equals b."

The letters a and b represent unknown numbers, but when you know what a is and add 2 to it, you can figure out what b is. For example, if you learn that the variable a is equal to 5, then the example becomes this:

$$5 + 2 = b$$

So the variable b is equal to 7.

Variables can be any letter you want, but x is a very popular variable name. You've probably heard teachers, students, and co-workers talk about "solving for x." Using x as a variable has a mysterious flavor, suggesting a great unknown. Think of *The X-Files*, "X marks the spot," and *Planet X* (the Star Trek novel).

Are your constants constant, Constance?

The opposite of a variable is a *constant*. It has a fixed value. For example, look at

$$a + 2 = b$$

Note that 2 is a constant. If variable a changes, it causes variable b to change, but the constant 2 stays the same.

Various types of numbers, such 3, 2.5, 1/2, and π (pi) are constants. Constants can be numbers of any kind.

Expressions and equations

After you understand the difference between variables and constants, you can begin to form variables and constants into expressions and equations.

Examining expressions

When life gives you lemons, make lemonade. When math gives you variables and constants, make expressions. An *expression* is a combination of symbols and can be made up of variables, constants, or both. That's it!

An expression isn't necessarily equal to anything; it's like using a phrase rather than a whole sentence. Here's a sample expression, made entirely of constants:

$$3 + 4 + 5$$

You can easily add these constants up.

Here's an expression made up entirely of variables:

$$a + b + c$$

In the example, you add some unknown quantity of something (a) to an unknown quantity of something else (b) and then add that to another unknown quantity (c).

You can group variables and constants together to form an expression. The most popular way to group constants and variables is to place them inside parentheses. For example, here's an expression with four variables, grouped by two sets of parentheses:

$$(a + b) + (c + d)$$

The value in grouping the items (also called *terms*) is that such groupings often make an expression easier to read and a problem easier to solve. And to increase flexibility, sometimes *ungrouping* items is useful.

Getting a handle on equations

An *equation* is similar to an expression in that they both are combinations of terms. The difference between an equation and an expression is that the equation has an equals sign (=). In an equation, the expression on the left side of the equals sign is equal to the expression on the right side.

For example, consider these two expressions:

$$3 + 4 + 7$$
$$5 + 8 + 1$$

These two expressions aren't much to write home about, but when you relate the expressions in an equation, things get slightly more exciting. For example, is the following equation true?

$$3 + 4 + 7 = 5 + 8 + 1$$

A little arithmetic on both sides of the equation gives you this:

$$14 = 14$$

What a relief! They are equal. Now in the world of algebra, equations containing variables are far more interesting. For example:

$$a + b = 14$$

You don't know what a and b are yet, but the equation declares that their sum is 14.

The equality relationship in an equation applies to all algebra, from the most trivial to the most sophisticated. Equations such as this come up a lot in simple word problems.

Operating with variables

You can do the same operations with variables that you do with numbers. You can add, subtract, multiply, and divide variables. You can also raise them to a power (called *exponentiation*) or find their square roots. You write variables just like numbers. (If you want a fancy name for this, the rules are called *mathematical notation*.) Here's what you need to know:

✔ **When you add and subtract variables:** You use the same notation you use to add and subtract numbers:

$a + b$

$a - b$

$a + 0$

$a - 0$

✔ **When you multiply variables:** You use special notation. You don't use the traditional times sign ($\times$) because it looks too much like the variable x. Instead, you use parentheses, a dot ($\cdot$), or no sign. For example, all of the following are ways to show that you're multiplying 2 and b.

$2 \cdot b$

$2(b)$

$2b$

✔ **When you divide variables:** You should never use the traditional division sign ($\div$). Instead, you express division as a fraction:

$\dfrac{b}{2}$

$\dfrac{2}{d}$

$\dfrac{x}{y}$

✔ **When you write powers and square roots of variables:** You write powers and square roots exactly the same way you'd write numbers:

f^2

$\sqrt{g}$

The following equation looks a little complicated, but it's composed entirely of simple variables and a constant. The equation just follows the rules of notation.

$$z = \frac{(h)\pi r^2}{27} c$$

The previous equation actually does something practical. In one algebraic "sentence," it shows how to calculate the volume of a concrete patio in square feet, divide to get cubic yards, and multiply by the cost per yard. The result is z, the cost of the concrete for the patio. (In pouring ready-mix concrete, *yard* is the term suppliers use for a cubic yard.)

You can do the same operations with variables that you do with numbers, and you need to follow the same rules of arithmetic for variables that you follow for numbers. See Chapter 1 for details on operations and the rules that apply to them.

Applying the same operation on both sides of the equal sign

When math in real life gives you a mash-up of facts, you can usually make a good equation. But a good equation is just the starting point. You need to solve it, too.

To solve an algebra problem, you must perform the same math operations on both sides of the equation. If you do so, the equation maintains its equality. That's essential for problem solving, and it works every time.

For example, here's an equation that tells you one thing:

$$a - 7 = b + 9$$

Add 7 to both sides. Because you're doing the same thing to both sides of the equation, you preserve the equality.

$$a - 7 + 7 = b + 9 + 7$$
$$a = b + 16$$

Now subtract b from both sides.

$$a = b + 16$$
$$a - b = b + 16 - b$$
$$a - b = 16$$

At this point, you have "cleaned up" the equation.

An essential tactic in solving many algebra problems is to get all variables on one side of the equation. In some cases, you want to get one variable on one side of the equation. Both techniques are valuable.

In the subtraction example, you don't know the final answer — yet — but you can see that $a - b = 16$. In the addition example, you are expressing the value of a in terms of a variable (b) and a constant (16).

This same technique works for other operations, too. In this next equation, you multiply both sides of an equation by 2. This isn't the simplest way to the solution, but it shows how performing the same operation on both sides of an equation preserves the equality:

$$a + 3 = 9$$
$$2(a + 3) = 2(9)$$
$$2(a + 3) = 18$$

Now to solve this equation, you can get rid of the parentheses on the left by multiplying each term in the parentheses by 2.

$$2(a + 3) = 18$$
$$2a + 6 = 18$$

To clean the equation up a bit more, subtract 6 from each side.

$$2a + 6 = 18$$
$$2a + 6 - 6 = 18 - 6$$
$$2a = 12$$

You can wrap things up by dividing each side of the equation by 2.

$$2a = 12$$
$$\frac{2a}{2} = \frac{12}{2}$$
$$a = 6$$

The answer is $a = 6$. Notice that through all the different steps, you preserve the equality by doing on one side of the equal sign whatever you did on the other side.

Keeping order with operations

When equations get complex, you solve them by doing operations in the correct order to simplify them. Not surprisingly, this correct order is called the *order of operations* (and sometimes it's known as *operator precedence*). Here's the order in which you perform the different operations in an equation, arranged from first to last:

✔ Terms inside parentheses or brackets

✔ Exponents and roots

✔ Multiplication and division

✔ Addition and subtraction

The rule of thumb is to work from the "inside out" (starting with terms inside parentheses) and make complex expressions (exponents and roots) simple.

For example, simplify

$$x = (5 - 3) + (22 + 6a) - (4a \cdot 3)$$

to become

$$x = 2 + 22 + 6a - 12a$$

Three-for-one equation bonus: Calculating speed, time, and distance

If you drive a car or have ever flown in an airplane, you've probably noticed that time, speed, and distance are related. Here's the basic formula for distance, based on speed and time:

$$distance = velocity \times time$$
$$d = vt$$

Distance equals speed multiplied by time. In science, the correct term for speed is *velocity*, represented by *v*. But wait! As they say on TV, there's more! The following related formulas are also true:

$$velocity = \frac{distance}{time}$$
$$v = \frac{d}{t}$$
$$time = \frac{distance}{velocity}$$
$$t = \frac{d}{v}$$

When you know two of the parts of the formula, you can solve for the third part. If, for example, you know the distance you've traveled and the time it has taken, you can calculate your average velocity. If you know the distance you've traveled and the average velocity, you can calculate the time you've been driving.

This is a great three-for-one bonus formula, and you'll find other examples in day-to-day math. For details, head to Chapter 10.

Fortunately, a good equation keeps things separate, with parentheses and math signs. Be careful, though, because a bad equation can be ambiguous. Don't blame yourself — unless you made up the equation.

Jousting with Geometry: Simple Rules about Shape and Size

The term *geometry* comes from the Greek words meaning "earth measurement" (even though some contend that it actually translates as "causes students pain and suffering"). But for the purposes of math you're most likely to use in real life, you can safely reduce the scope of geometry from measuring the earth to measuring for a pool, deck, or patio, or for laying out a playing field for soccer, badminton, or volleyball.

Geometry goes back a long way, at least to ancient Egypt and Babylonia. It makes sense that when the ruler wanted tax money from farmers, the process began by measuring the farmers' fields. That's where geometry comes in. Euclid, the Greek mathematician, gets the credit for giving us formal geometry. He developed principles in about 300 BCE, and so *Euclidian geometry* is the kind of geometry you learned in school. It's very abstract and is based on axioms and proofs. Euclidean geometry is fascinating stuff, but it's not very practical for day-to-day problems.

Looking at geometry's basic parts: Planes, points, and lines

The geometry used in everyday life is *plane geometry* and *solid geometry*. Plane geometry is a world of points, lines, and shapes — all of which take place on a plane. Solid geometry is a world of volumes.

Plain talk about the plane

A *plane* is a flat, two-dimensional surface. Being theoretical, a geometric plane is perfectly flat and extends forever in all directions. In real life, you draw geometric figures on a flat piece of paper or a flat computer screen. If you're lucky, your lawn is fairly flat, and your street is, too.

To make geometry work, there must be a *coordinate system*, which is a way of describing the position of any object on a plane.

The most famous and commonly used coordinate system is the *Cartesian coordinate system* (named after René Descartes, the famous French mathematician and philosopher). Figure 2-1 shows the Cartesian coordinate system.

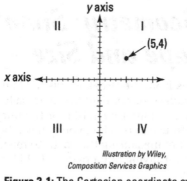

Illustration by Wiley,
Composition Services Graphics

Figure 2-1: The Cartesian coordinate system.

The system has two axes (plural for *axis*). The horizontal axis is the *x*-axis, and the vertical axis is the *y*-axis. Along each axis are points, and the two axes cross each other at point (0,0), called the *origin*. As a bonus, you get four quadrants, named I, II, III, and IV.

You can describe any position on the plane by naming coordinates. In Figure 2-1, the point shown is at coordinate (5,4). Its location is 5 to the right of the origin and 4 up from the origin. The pair of numbers describing a point's position on the plane is called an *ordered pair*.

Getting to the point

The *point* is the basic building block of geometry. In theory, it has no height or width. In real life, it's about the size of a pencil point or the little hole that a pin makes.

Each point represents a place on the plane. It's a precise location. If the GPS map on your smartphone is a plane, then the blue dot (that's what I've got on my phone) is the point where you are.

In fact, all reading and following of maps amounts to plotting points (your locations) on a plane. With the smartphone or a GPS, the application does it for you. With a paper map, you do it manually.

Falling in line

If the point is the basic building block of geometry, then the line is the next step. It should be called "Son of Point, the Sequel." A line is straight, theoretically has length but no width, and continues on in either direction without stopping.

Walking in a city with a grid layout

Many cities are laid out in a grid pattern. They often have numbered streets (10th Street, 11th Street, and so on) going in one direction, and lettered streets (A Street, B Street, and so on) going in the other. Some have numbered avenues instead.

Say you're in a city with a grid, and you want to get from Point A to Point B. You find out that Point B is three blocks east and two blocks north. Here are a couple of ways to get there:

- Think of point A as (0,0). Go east three blocks. You are now at (3,0). Go north two blocks. You have now arrived at Point B (3,2).

- Alternatively, you could go north two blocks, arriving at (0,2). Then you go east three blocks, putting you at (3,2). The route is different, but the result is the same.

This technique is common in cities with a grid, and knowing it is handy if you need to ask for directions. "Yeah, just go east three blocks and north two blocks. You can't miss it." Well, if directions such as north, south, east, and west give you trouble, make sure the direction-giver points which way to go.

A straight line is the shortest distance between two points, which brings us to the discussion of line segments. *Line segments* are lines that have a beginning and ending point. To be formal, a line segment can be represented by its two end points. Figure 2-2 shows a line with end points (0,0) and (5,0).

(0,0)   (5,0)

Illustration by Wiley,
Composition Services Graphics

Figure 2-2: A line segment.

The line is common in life. It has hundreds of uses and has populated our vocabulary. Think "walk the line," "draw the line," and "that's a line if I ever heard one."

If you like to spend your evenings working out first-degree polynomial functions of one variable, you'll see a lot of lines. But if that's not useful or fun, you also use lines in laying out fencing for your house or for drying clothes on a clothesline. They're the same thing.

Thanks to the line and the point, you can do some excellent things with angles, shapes, areas, and volumes.

What's your angle? Acute, obtuse, right angles, and more

Geometry includes a lot of angles, and you should know them. An *angle* is a geometric figure made up of two lines, joined at an end point called the *vertex*. In an angle, the lines go on forever and are known as *rays*.

Angles come in just a few flavors, based on how wide or narrow they are. Angles are measured in degrees. The smallest angle is theoretically 0 degrees, but that's very boring. The largest angle is 360 degrees, which is a full sweep. That's what 12:00 PM looks like on a clock. Figure 2-3 shows various types of angles.

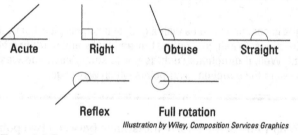

Acute   Right   Obtuse   Straight

Reflex   Full rotation

Illustration by Wiley, Composition Services Graphics

Figure 2-3: Types of angles.

Here are angle basics:

- ✔ **Acute:** This angle is less than 90 degrees.

- ✔ **Right:** This angle is exactly 90 degrees.

- ✔ **Obtuse:** This angle is greater than 90 degrees but less than 180 degrees.

- ✔ **Straight:** This angle is 180 degrees and doesn't look much like an angle.

- ✔ **Reflex:** This angle is greater than 180 degrees but less than 360 degrees.

- ✔ **Full rotation:** This angle is 360 degrees and doesn't look much like an angle.

You should know your angles in order to communicate with others. In some cases, you actually use angles.

- ✔ If you're involved with crafts or woodworking, you probably need to cut things at angles.

 ✔ If you go to the picture framing shop, you can ask the salesperson what angle the bevel-cut matte will be.

 ✔ When you talk to the carpet layer, you can describe an odd-shaped room's angles.

 ✔ If you help your kids with math, you'll look smart and save yourself embarrassment.

The shape of things

The world is filled with fascinating shapes, both beautiful and practical. Although many shapes exist, life is short, so this section describes only three of them — rectangles, triangles, and circles, which just happen to be the ones you'll use most often in your real-life math problems.

Rectangles and squares

A *rectangle* is a four-sided figure. It has a length and a width, and each corner is a right angle. A *square* is a special rectangle. The length and width are the same. Figure 2-4 shows a square and a rectangle.

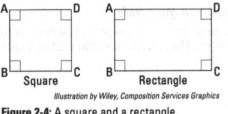

Illustration by Wiley, Composition Services Graphics

Figure 2-4: A square and a rectangle.

Triangles

A *triangle* is a three-sided figure. Triangles come in several flavors, and each one has its own name.

Figure 2-5 shows different types of triangles. Here are triangle basics:

 ✔ **Acute triangle:** The angles of an acute triangle are all less than 90 degrees.

 ✔ **Right triangle:** A right triangle has one angle of exactly 90 degrees.

 ✔ **Obtuse triangle:** An obtuse triangle has one angle greater than 90 degrees.

- **Isosceles triangle:** An isosceles triangle has two sides of equal length and, therefore, two angles with the same value.

- **Equilateral triangle:** All three sides of an equilateral triangle are equal in length; therefore, the three angles have the same value.

- **Scalene triangle:** A scalene triangle has sides of three different lengths and therefore three different angles.

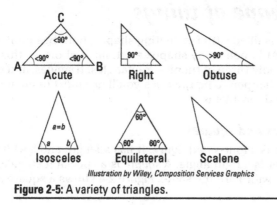

Illustration by Wiley, Composition Services Graphics

Figure 2-5: A variety of triangles.

'Round and 'round you go: Circles

A *circle* is a shape in which every point on the edge is the same distance from the center. That distance is the radius. Figure 2-6 shows the parts of a circle.

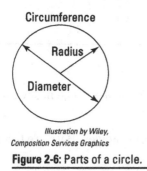

Illustration by Wiley,
Composition Services Graphics

Figure 2-6: Parts of a circle.

Here's what you need to know about circles:

- **Radius:** The radius is the distance from the center to the edge, known in formulas as *r*.

- **Diameter:** The diameter is the distance across the circle, through the center. The diameter (also called *d*) is equal to 2 times the radius, or 2*r*.

✔ **Circumference:** The circumference is the distance around a circle, known in formulas as *c*.

Calculating areas

An *area* is a quantity of two-dimensional space. You need to know about three areas: areas of the rectangle, triangle, and circle. Fortunately, the area formulas you're likely to use are both easy to understand and easy to use. By the way, it doesn't matter whether you're talking about square inches or square miles; the formulas are the same.

Finding the area of squares and rectangles

The formula for calculating the area of a rectangle is very simple: You simply multiply the rectangle's length by its width. Figure 2-7 shows a rectangle's length and width.

Length

Illustration by Wiley, Composition Services Graphics

Figure 2-7: The length and width of a rectangle.

The official formula looks like this:

$$area = length \times width$$
$$a = lw$$

Finding the area of triangles

The area of a triangle is also easy to calculate. Every triangle has a base and a height. Figure 2-8 shows a triangle's base and height.

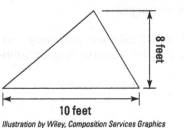

10 feet

Illustration by Wiley, Composition Services Graphics

Figure 2-8: A triangle's base and height.

The formula for calculating the area of a triangle is also very simple. Just multiply the base by the height and divide by 2:

$$\text{area} = \frac{\text{base} \times \text{height}}{2}$$

$$a = \frac{bh}{2}$$

Finding the area of circles

The area of a circle is easy, too. Use the radius of the circle, shown in Figure 2-9, and the "magic" number pi (π), which is approximately 3.14159.

Illustration by Wiley,
Composition Services
Graphics

Figure 2-9: A circle's radius.

To find the area of a circle, use this simple formula, in which you multiply π by the radius squared:

$$a = \pi r^2$$

Getting pushed to the edge: Perimeters

The distance around a geometric figure, such as a square, rectangle, or triangle, is called the *perimeter*. The word comes from the Greek *peri* (around) and *meter* (measure). In the case of a circle, the perimeter gets a special name — *circumference*. Figure 2-10 shows the perimeter of a rectangle.

Knowing a perimeter is handy (and some say essential) for measuring fencing, a paddock, a circular exercise ring, or bender boards for flower beds.

Illustration by Wiley,
Composition Services
Graphics

Figure 2-10: The perimeter of a rectangle.

You can find a rectangle's perimeter, using one of the following methods:

- ✔ Measure each side and add the numbers up.
- ✔ Double the length and double the width and add them together. The formula is

  $p = 2l + 2w$

- ✔ In the special case of a square, where all four sides are equal, just multiply the length of one side by 4.

To find the perimeter of a circle, use the famous (but common) circumference formula, in which you multiply π by the circle's diameter:

$c = \pi d$

Speaking volumes about boxes

Volume is a quantity of three-dimensional space. That space has length, width, and height. The official name for a box-shaped item is *cuboid*. The unofficial name is *box*. Figure 2-11 shows a cuboid.

Think of volume as a measure of "how much of something" you are buying or using. While some items are measured by length or weight, volume is the standard measure for almost all liquids.

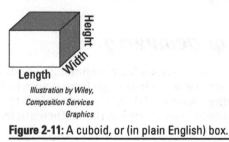

Illustration by Wiley,
Composition Services
Graphics

Figure 2-11: A cuboid, or (in plain English) box.

A math secret

The liter and the milliliter are very important, even in the non-metric United States. Medicine and science rely on these units. The nurse or scientist usually dispenses liquids, and the liter and milliliter are actually measures of cubic *volume*, based on length, width, and height. One liter (1 L) is the *volume* of a cube with dimensions of 10 centimeters (10 cm) on each side. One milliliter (1 mL) is one-thousandth of a liter, which is the *volume* of a cube with dimensions of one centimeter (1 cm) on each side.

The common units of volume are cubic inches or cubic centimeters (which you'd want to know to determine engine displacement), cubic feet or liters (good for comparing refrigerator capacities), and cubic yards or cubic meters (the units that specify how much ready-mix concrete you need).

In addition, we buy many items around us by liquid measure (a 1-gallon jug, for example). The gas tank in your car is described by its capacity in gallons or liters. Even the storage lugs in your garage have capacities labeled in gallons or liters. Firewood is sold by the cord (which, interestingly, is defined *by law* in most states of the U.S. and typically refers to a stack of wood 4 feet wide, 4 feet high, and 8 feet long — 128 cubic feet).

As part of your brief education or review of geometry, you should be able to calculate the volume of a cuboid. The formula is very simple:

volume = length × width × height

$v = lwh$

To see a real-life example of a cuboid, go to one of the popular rent-a-truck moving companies. Ask for a book box. Its dimensions are 12 inches x 12 inches x 12 inches. Because 12 inches are in a foot, that's 1 foot x 1 foot x 1 foot, giving a volume of 1.0 cubic foot (because $1 \times 1 \times 1 = 1$).

Summing up geometry

If you read the preceding sections from beginning to end, you may have noticed that I arranged the info to go from points to lines to flat shapes (rectangles, squares, triangles, and circles) to cuboids. Those items fit in *dimensional space* (as the mathematicians say) in an orderly way.

Table 2-1 summarizes the common geometric shapes out in the world, showing the name, the dimension they "live in," and what you measure or calculate.

Table 2-1	Geometric Shapes	
Name	**Dimension**	**Measurement**
Point	0	Position
Line	1	Length
Rectangle, square, triangle, circle	2	Area
Cuboid	3	Volume

Chapter 3

Becoming a Believer: Conversion, Statistics, Probability, and More

- -

In This Chapter

▶ Using the ratio-proportion formula

▶ Understanding conversions

▶ Interpreting basic statistics

▶ Knowing your chances through probability

- -

*T*he math problems in real life take many forms. Although these problems can be a challenge, their solutions take only a few forms. Basic arithmetic (the topic of Chapter 1) handles a lot, from figuring mortgage interest to calculating cantaloupes. And basic algebra (which you can read about in Chapter 2) takes care of most of the rest.

Then the basics evolve into specialties like statistics, conversions, and probability. Fortunately, the real-life math you'll encounter in these areas involves math you already know how to do. For example, two common statistics are based on addition, division, and counting, and unit conversions are based on multiplication and division.

The same is true of probability. The math for basic calculations is simple. You need only to know some terms and how to apply them.

In this chapter, I explain the basic principles of conversion, statistics, and probability, and show you the greatest formula ever, which is almost a universal problem solver. As a bonus, I give you a brief rundown of the best tools to use to handle real-life math.

Wrangling Ratio-Proportion: The Best Calculation Method

If 1 (of something) gets you 2 (of something else), then 2 gets you 4. That, in a nutshell, is the entire concept of ratio-proportion.

To understand, you first need to know what a ratio is. A *ratio* is the relationship between two quantities. Some ratios are obvious. For example, if 8 slices are in 1 pie, the ratio is 8 slices per pie and it's presented mathematically like this:

$$\frac{a}{b}$$

Ratio-proportion is simply a calculation method that compares two ratios. The two ratios amount to four items. If you know three of them, you can solve for the fourth.

A *proportion* is the relationship between four quantities, shown in the following equation. You say this equation as "*a* is to *b* as *c* is to *d*." The first item divided by the second is equal to the third item divided by the fourth.

$$\frac{a}{b} = \frac{c}{d}$$

A proportion doesn't change. For example, if 1 cup of flour produces 8 pancakes, the proportion is the same whether you're using 100 or 1,000 cups of flour.

A ratio-proportion equation has a ratio on the left that's equal to another ratio on the right. Going back to the 8 slices in a pie example, a ratio-proportion question might ask how many slices are in 2 pies. This one's easy (as pie). There are 16 slices in 2 pies. Here's the structure for solving a ratio-proportion problem, using the pie example:

$$\frac{\text{known equivalent}}{\text{known equivalent}} = \frac{\text{known equivalent}}{\text{desired equivalent}}$$

$$\frac{1 \text{ pie}}{8 \text{ slices}} = \frac{2 \text{ pies}}{x \text{ slices}}$$

$$x = 2 \times 8$$

$$x = 16$$

The math amounts to multiplying *a* (1 pie) by *d* (x slices) and *b* (8 slices) by *c* (2 pie), in a maneuver called *cross-multiplying*.

You can also flip the both sides upside down, but you still have to cross-multiply:

$$\frac{\text{known equivalent}}{\text{known equivalent}} = \frac{\text{desired equivalent}}{\text{known equivalent}}$$

$$\frac{8 \text{ slices}}{1 \text{ pie}} = \frac{x \text{ slices}}{2 \text{ pies}}$$

$$2 \times 8 = x$$

$$16 = x$$

Try out ratio-proportion. Say you have 2 bags of apples with 6 apples in each bag. Your crazy uncle just gave you 81 bags. How many apples did he give you?

1. Set up the problem.

$$\frac{\text{known quantity}}{\text{known quantity}} = \frac{\text{known quantity}}{\text{desired quantity}}$$

$$\frac{2 \text{ bags}}{12 \text{ apples}} = \frac{81 \text{ bags}}{\text{total number of apples}}$$

2. Cross multiply to solve.

$$\frac{2}{12} = \frac{81}{x}$$

$$2x = 12 \times 81$$

$$2x = 972$$

$$x = 486$$

The answer is 486 apples.

Doing Conversions: Lots of Pleasure and Hardly Any Pain

The world is the product of thousands of years of civilization. That's good. But those millennia have produced many different systems of measurement. That's bad.

The situation is worse when you need to convert quantities (like distance, weight, and volume) from one measurement system to another. It's especially difficult in the United States, because the main system in the U.S. is American units, while the rest of the world uses the metric system.

No problem! With a little understanding and the right tools, you will be a conversion whiz. Understanding takes a little effort, and the tools are free.

Factoring in the conversion factor

A *conversion factor* is a simple formula that lets you convert from one unit to another. Conversion factors are fast, fun, friendly, and common in everyday life. You probably know a bunch of them already. For example, here are a couple of conversion factors: Twenty-four hours are in a day, and 12 inches are in a foot.

For conversion factors you don't know, here's a simple math secret: To convert from one unit to another, you either multiply or divide. You just need to know what value to multiply or divide by.

A simple example is converting feet to yards. Say you want to know how many yards are in 81 feet. You know a yard has 3 feet. So you divide 81 feet by 3 to get yards. Here's the equation:

$$\text{Yards} = \frac{\text{feet}}{3}$$
$$\text{Yards} = \frac{81}{3}$$
$$\text{Yards} = 27$$

To find any conversion factor you don't already know, go to http://www.google.com and enter the conversion you want. For example, enter "feet to miles," "tons to pounds," and so forth. Almost everything produces an immediate display on the "presearch," and you only have to click on that.

Using United States customary units

United States customary units (the "American system") are measurements used in the United States. The basic units are

- ✔ **Length:** inch, foot, yard, mile
- ✔ **Area:** acre, square foot
- ✔ **Volume:** cubic inch, cubic foot, cubic yard
- ✔ **Liquid volume:** fluid ounce, pint, quart, gallon, teaspoon, tablespoon, cup
- ✔ **Weight (mass):** ounce, pound, ton

The easiest way to convert between units of the American system or from American to metric is to use an Internet calculator.

You can display the conversion factor online with no effort. Go to http://www.google.com and enter the conversion you want.

The preceding has more details on using online conversion calculators. (If you're working in the kitchen — where most real-life conversion conundrums occur — head to Chapter 6 for a handy-dandy conversion chart.)

Managing the metric system

The *metric system* is officially called the International System of Units. Most of the world, except for three countries (the United States, Liberia, and Burma/Myanmar), uses the metric system. The basic units in the metric system are

✔ **Length:** meter, kilometer

✔ **Area:** hectare

✔ **Volume:** liter

✔ **Weight (mass):** kilogram, tonne

Metric units are easy, because every unit is a multiple of ten of another unit. The easiest ways to convert between metric units or from metric to American units is to use an Internet calculator. You can also display conversion factors online with no effort. See the earlier section, "Factoring in the conversion factor" for details.

If you work in medicine or science, you're practically home free, because you work with metric units all the time.

 Temperature is expressed in degrees Fahrenheit in the United States and in degrees Celsius in the rest of the world. To convert from Fahrenheit to Celsius, use this formula:

$$°C = (°F - 32) \times 5/9$$

To convert from Celsius to Fahrenheit, use this formula:

$$°F = °C \times 9/5 + 32$$

Mastering Simple Statistics

Statistics is the study of data. A person collects data, organizes it, draws inferences from it, and presents it (usually in tables or charts). You can be sure that businesses, scientists, government agencies, medical researchers, and economists create and use statistics a lot. And there's a specialty: A professional number cruncher is a *statistician*.

Statistical information (*stats*) can be your friend. The world's filled with uncertainty, but statistics help you make decisions in the face of uncertainty.

For example, many people want to see how home prices or home mortgage interest rates are trending. The trend might give you an indication of when to buy if you want to save big money. Similarly, watching how your salary increases are trending can help you draw some conclusions. If it's trending at about 1 percent per year (not good!), you know you probably aren't going to see a big difference next year or the year after. In the following sections, I tell what you need to know about statistics to make educated assessments of data.

Don't use statistics blindly. The old joke is "Did you hear about the statistician who drowned in a river with an average depth of 3 feet?"

An average is a mean thing

An *average* (also known as an *arithmetic mean*) is the result of adding up numbers in a collection and dividing by the number of items in the collection. Here's the equation:

$$\text{Average} = \frac{\text{sum of items}}{\text{number of items}}$$

The nice thing about averages in real life is that you see a lot of them. Two common averages are average bowling score and grade point average (GPA). Of course, you can average just about anything, if the results will be meaningful to you. For example, you might average several weeks of grocery purchases to get a broad view of what you're spending (an exercise that's great when you're trying to predict spending when you create a budget, as I explain in Chapter 10). To get your weekly average grocery bill, you would follow these steps:

1. **Add the total purchases.**

   Total = $45.00 + $50.00 + $45.00 + $125.00 + $20.00

   Total = $285.00

2. **Divide by the number of times you bought groceries.**

   $$\text{Average} = \frac{\$285.00}{5}$$

   Average = $57.00

One week you spent $125.00 (maybe you had a big family dinner). The next week you spent far less (maybe you were eating leftovers), and the other amounts are all pretty similar (between $45 and $50). But taken all together, the average is $57.00.

Mediating the median

The median is the number that separates the upper half of a sample from the lower half. A median is not the same as an average. You simply arrange the values in order from lowest to highest and find the number in the middle of the list. That's your median value. If the list has an even number of items, you average the two middle numbers.

You can easily find some medians. Say you want to find the median prices of the eight homes that recently sold in your neighborhood. Follow these steps:

1. **Record the prices of the recent sales.**

   Say the houses sold for these amounts:

   > $196,000; $175,000; $190,000; $199,000; $220,000: $193,000; $187,000; and $195,000

2. **Arrange the values in order from lowest to highest.**

   > $175,000; $187,000; $190,000; $193,000; $195,000; $196,000; $199,000; $220,000

3. **Count halfway through the items.**

   The number in the middle is the median. If the list has an even number of items (as it does here), average the two middle members, which in this case are $193,000 and $195,000. The median price of the recently sold homes is $194,000.

Statisticians like medians, because unlike averages, medians usually don't get distorted by *outliers*. Averages can be misleading if you don't know what to look for. For example, the median 2010 price of a home in the United States was $221,800, but the average price was $272,900. Here, the average price is higher than the median price, and the likely cause is that a small number of very expensive homes raised it. Remember, you interpret the median to mean that half the homes had prices of less than $221,800, and half had greater prices.

Medians work best with large samples and a fairly even distribution of values.

Figuring percentiles

A *percentile* is a number between 0 and 100. It's a value below which a certain percent of scores fall. For example, if you score at the 70th percentile of a test, your score is greater than 70 percent of other people taking the test. If you score at the 50th percentile, that's the right in the middle.

The 25th percentile is also called *first quartile* (Q1), the 50th percentile is the *second quartile* (Q2), and the 75th percentile is the *third quartile* (Q3).

You can calculate test percentiles yourself, provided you know the number of people taking the test and their scores. Say 30 people take a test. You scored 89 out of 100 points. You see that 24 of the 30 test takers (which is 80 percent of them) scored below 89. Your score is greater than 80 percent of the other people, putting you in the 80th percentile.

You tend to read percentiles more than calculate them. If you visit the U.S. Bureau of Labor Statistic and check any career (your current one or maybe one you'd like to get into), you'll see some salary percentiles.

Being aware of statistical fallacies

Beware of misleading statistics, which can occur in business, government, and politics. Benjamin Disraeli, the great British prime minister, is supposed to have said, "There are three kinds of lies: lies, damned lies, and statistics." Often, the problem is a false implication that's accidental. But sometimes statistics seem designed to be misleading. Keep your analytical thinking cap on!

Are you SATisfied?

On some tests, such as the SAT, the raw score is the number of points gained from correct answers less the number of points lost from incorrect answers.

The SAT has three parts, each with a score value of 200 to 800 points, and a maximum score of 2,400 points. In 2011, the average score in the United States was 1500: 489 for writing, 514 for math, and 497 for critical reading.

Bluntly, these average scores probably won't get you admitted to a first-class college. You need to score at the 90th percentile or above. To score at the impressive 98th percentile, try for a score of 2250: 750 for writing, 760 for math, and 750 for critical reading.

But don't despair if your SAT score isn't quite what you want it to be. In the admission process, colleges also consider GPA, extracurricular activities, and other factors.

Bogus statistics can produce major problems in decision-making and could possibly cause disastrous consequences. Distortions come in many forms. Here are four:

✔ **Misuse of rate-of-change data:** Rate of change isn't change. Rate of change can make it appear that some declines (such as unemployment) and some increases (such as salary) are big when they aren't. If your salary went up 1 percent last year and 2 percent this year, for example, that's change — a change of 1 percent and 2 percent. The distortion comes if your boss says "Your salary increase *doubled*, compared to last year." Yes, 2 percent is double 1 percent and is mathematically correct, but that's not a big deal.

✔ **Distorted visual representations:** Charts can contain distortions. For example, Figure 3-1 shows a proposed 4.6 point increase in the highest income tax rate (from 35 percent to 39.6) in a way that makes the change appear to be much larger than it is. If the chart showed the entire length of the bars (from 0 to 35 and 0 to 39.5), the visual would show that the increase isn't dramatic. Instead, the chart seems to show that the second bar is *6 times* the height of the first bar. Such distortions are sometimes accidental but are frequently deliberate.

✔ **Cherry-picking and using meaningless stats:** Beware of stats that either don't answer the question asked or that try to put a good spin on bad news. For example, imagine a school claiming a graduation rate of over 90 percent. Sounds great, right? Now imagine that the school, in doing its calculations, didn't count a large number of students who had dropped out.

✔ **Distortions in your mind:** Most statistics apply to large groups. They don't apply to individual cases, so don't let your mind create a distortion. For example, if you read that 10 percent of people in a certain group will die of cancer, *don't* count out ten friends and convince them that one of your group is at death's door.

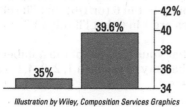

Illustration by Wiley, Composition Services Graphics

Figure 3-1: Distorted visual representation.

TIP

Always feel free to question statistics. Granted, you don't have time to prove them all, but stay alert.

Predicting the Probable

Many events can't be predicted with certainty, so people look at the *likelihood* that events will occur. Enter probability. *Probability*, at its simplest, is the number of cases where an event will happen compared to the total number of possible outcomes.

Probability is essential to the insurance and gaming (gambling) industries. It's also used in medicine, science, and engineering — wherever there are complex systems and only partial knowledge.

Most often, you are a consumer of probabilities that someone else develops. However, you can always do your own probability experiments and draw your own conclusions from the results.

Determining probability

As stated, probability compares the cases where an event will happen to the total number of possible outcomes. Probability is a value between 0 (it will never happen) and 1 (it's an absolute certainty).

The sun has such a good record for rising in the morning that you can safely assign the event a probability of 1. I have such a poor likelihood of becoming a rock star that you can safely assign the event a probably of 0. Other events fall somewhere in between. For example, a die has six sides with dots (*pips*) representing 1, 2, 3, 4, 5, and 6. Using the formula for probability, the chances of rolling a 1 are

$$p = \frac{1}{6}$$

The formula looks exactly like a fraction or division problem. You says this as "The probability is 1 in 6 (or 1/6) that I'll roll a 1." You can also say, "My chances are 1 in 6 that I'll roll a 1."

By the way, no matter what your roll, one of the numbers *will* come up. To see the probability of rolling a 1, 2, 3, 4, 5, or 6, add the individual probabilities.

$$p = \frac{1}{6} + \frac{1}{6} + \frac{1}{6} + \frac{1}{6} + \frac{1}{6} + \frac{1}{6}$$

$$p = \frac{6}{6}$$

$$p = 1$$

Here's another example. Suppose you have 20 socks in a drawer and only 2 of them are red. What's the probability of pulling out a red sock? Compare the number of desirable outcomes (getting a red sock) to the total number of outcomes:

$$p = \frac{2}{20}$$
$$p = \frac{1}{10}$$

There's 1 chance in 10 that you'll pull out a red sock.

What's the probability of pulling out 2 red socks in a row? Keep in mind that pulling out the second red sock is an independent event that has nothing to do with pulling out the first red sock. The only consequence of pulling out the first sock is that now you now have only 19 socks in the drawer.

The probability of two favorable events (getting the first red sock and then getting the second) is the *product* of their two probabilities — 1 chance in 10 for the first sock and 1 chance in 19 for the second sock:

$$p = \frac{1}{10} \times \frac{1}{19}$$
$$p = \frac{1}{190}$$

The chances are 1 in 190 of your pulling out 2 red socks in a row.

Here's one last example, and it's a classic: What's the probability that a coin will come up heads or tails?

First, you know that, if you flip a coin, it will come up on one side or the other. So you figure the probability of the coin coming up heads (or tails) by using this formula:

$$p = \frac{1}{2}$$

What happens if the coin comes up heads 1,000 times in a row? Does that mean you're due to see tails, or is the trend toward more heads? In this case, the probability of the coin coming up heads is

$$p = \frac{1}{2}$$

The probability is exactly the same every time. Independent events don't "influence" each other.

What are the odds?

Odds are the ratio of a favorable outcome to an unfavorable
outcome. Odds are essentially the same as probability but
expressed differently. In rolling a die, the probability of rolling
a 1 is 1 chance in 6, or 1/6. There's 1 favorable outcome and
5 unfavorable outcomes. You would express the odds in this
scenario as 5:1, and you'd say it as "5 to 1 against." To find out
more about odds, head to Chapter 9.

Chapter 4

The Miracle of Mental Math

● ●

In This Chapter

▶ Adding, subtracting, multiplying, and dividing large numbers

▶ Making reasonable estimates

▶ Looking at common statistics — averages and median values

● ●

*T*he handiest calculator is between your ears. Your brain is fast at doing math and doesn't have any batteries to recharge. In day-to-day activities, mainly shopping, mental mathematics (being able to solve math problems without paper, a pencil, a calculator, or a computer) is very convenient.

In this Chapter, I give you pointers on how to improve your mental math capabilities and offer scenarios featuring common math problems that you can solve just by working things out in your head. I also cover estimating, a trick anyone who's ever been shopping has used at one time or another, and simple statistics, just because they're interesting and fun.

Mental Math Basics

The secret to success with mental math is to tackle simple problems that have good prospects for success. Although in time you could learn to solve very complex problems, doing so is more of a parlor trick than a useful skill.

Follow this general approach to mental math:

✔ **Memorize some numbers.** Which numbers? These:

- **Your multiplication table.** That's not difficult, because you probably learned the table in the third grade.

- **Common equivalencies.** The number of minutes in an hour or feet in a mile (or, alternatively, meters in a kilometer) is useful, too.

✓ **Know some arithmetic basics.** Get familiar with the common operations (additions, subtraction, multiplication, and division). Often, a complex looking problem becomes a simple problem if you know how to break things down. For example, many "complex" multiplication problems are simple to solve if you can break them down into "multiply and then subtract" problems. The same is true of breaking up complex-looking addition problems.

✓ **Know when to stop.** Find your personal limit (in interest and skill) to doing certain operations. For example, I can multiply a four-digit number by a two-digit number in my head, but that's it. For anything bigger, I'm going to the calculator.

Adding and Subtracting on the Fly

Quick addition and subtraction have one simple rule: Break the problem into parts. As you see in the following section, you can usually represent each number in an addition or subtraction problem as two smaller numbers that are easier to work with.

One key to making this technique work is to think about problems a bit differently than you did in school. For example, you learned to add and subtract from the right, starting with the ones column. In mental math, you get faster results by working from the left, starting with the leftmost column.

Adding numbers quickly

To quickly add two numbers, even long ones, start at the left, at a column that seems comfortable to you. Break out the "easy" part and the "hard" part.

For example, say you want to add 2,344 and 698. Break the hundreds out from the tens and ones. The equation looks like this:

$$\text{total} = 2,344 + 698$$
$$\text{total} = (2,300 + 44) + (600 + 98)$$

Adding 2,300 and 600 is fairly easy; in fact it's just like adding 23 and 6. The answer is 2,900.

Now, handle those other pesky items, the 44 and 98. A good technique is to make 98 into 100 by reducing the 44 by two. The equation looks like this:

$$\text{total} = 44 + 98$$
$$\text{total} = 42 + 100$$

This little piece of "surgery" gives you a nice round 100 and a very manageable 42. Add the 100 to 2,900, giving 3,000. Now tack on the 42. The answer is 3042.

For a faster solution, you could just take 2 from the 44 and turn 698 into 700. The object is always to make "easy" numbers out of "hard" numbers.

Subtracting numbers quickly

To subtract two numbers, you use the same technique you use to add: You start at the left, at a column that seems comfortable to you, and break out the "easy" part and the "hard" part.

For example, say you want to subtract 530 and from 2,908. Break the hundreds from the tens and ones and then do the math on the resulting easier problems. The equation looks like this:

$$\text{difference} = 2,908 - 530$$
$$\text{difference} = (2,900 + 8) - (500 + 30)$$
$$\text{difference} = (2,900 - 500) + (8 - 30)$$
$$\text{difference} = 2,400 + 8 - 30$$

At first, you're off to an easy start. Take 500 from 2,900 and you're left with 2,400.

That leaves taking 30 from 8. Oops! Oh, my! What should you do? Answer: "Grab back" 100 from your partial result of 2,400 (leaving it at 2,300) and "give it" to the 8. Now you have 108 – 30. Subtract 30 from 108, and you're left with 78.

Adding 2,300 and 78 is no problem. The answer is 2,378.

Making Hay of Multiplication and Division

Like addition and subtraction, multiplication and division are easy to do in your head when you alter the numbers slightly to get a speedy result. Apply your knowledge of the multiplication table whenever possible.

Multiplying in your head

Multiplication is easy when the numbers are nice and round, but it can quickly get overwhelming when they aren't. For example, if four items are $5.00 each, you can pretty easily determine the total, because you know that $4 \times 5 = 20$. Ten items at $5.00 each will be $50.00.

But what if the numbers aren't nice and round? The secret is to round up to a number that's easy. Then subtract the small difference between the rounded number and the actual starting number, perform the multiplication on this small difference amount, and then do a little subtraction. (Trust me: It sounds more complicated that it is!)

Suppose, for example, that four items are $4.97 each and you want to find the total.

This problem, if you're using traditional multiplication, would require a calculator for many of us:

$4 \times \$4.97 = \$19.88$

But you can do it mentally by first rounding up from $4.97 to $5.00 and multiplying. The result is $20.00 ($4 \times \$5.00$). Then you take the difference between the two numbers ($0.03) and multiply, giving you $0.12 ($4 \times \$0.03$). Last, you subtract the multiplied difference from the multiplied rounded amount.

First, multiply the "big" rounded amount.

$4 \times \$5.00 = \$20.00$

Notice that you left off $0.03 per item? That's okay. Multiply the "small" difference amount separately.

$4 \times \$0.03 = \$0.12$

Now, simply take the $20.00 and subtract the $0.12. The answer is $19.88.

You deserve a break today! Here are some easy, handy, and fast multiplication techniques.

 ✔ Multiplying by 10 requires only that you append a zero to the end and move the decimal one place to the right. For example, if you multiply $5.00 by 10, you get $50.00.

> ✔ Multiplying by 5 is almost as easy as multiplying by 10. Just multiply the quantity by 10 (append a zero and move the decimal point to the right) and then divide the result by 2. To determine what $15 × 5 is, for example, multiply $15 by 10 ($15.00 × 10 = $150.00) and divide by 2 ($150.00 ÷ 2 = $75.00). The answer is $75.

Dividing in your head

Dividing in your head appears to be a little more complicated than multiplying, but appearances can be deceiving. Dividing is simple. The general idea is to divide the "big" division problem into smaller "little" division problems.

Say you want to divide 128 by 4. Fortunately, dividing by 4 is quite easy.

$$\text{result} = \frac{128}{4}$$
$$\text{result} = \frac{120}{4} + \frac{8}{4}$$
$$\text{result} = 30 + 2$$
$$\text{result} = 32$$

How do you get the answer? Find the first part of 128 that can be divided by 4. That's 12, the first two digits of the number 128. Then divide. The answer is 3.

Next, look at the rest of 128, which is 8. It easily divides by 4, giving you 2.

You may be asking yourself, "What if the numbers aren't all nice and round?" Not to worry. The principle is the same. For example, if you need to divide 131 by 4, just expand your equation a tad.

$$\text{result} = \frac{131}{4}$$
$$\text{result} = \frac{120}{4} + \frac{8}{4} + \frac{3}{4}$$
$$\text{result} = 30 + 2 + \frac{3}{4}$$
$$\text{result} = 32\frac{3}{4}$$

Divide everything that comes out "even." What's left is a proper fraction. Then do another division to get a decimal fraction. In the example, 3/4 = 0.75, so the result of the division is 32.75. When you apply this principle to money, the result is $32.75.

The fastest, easiest division is when you're dividing by 10 or 5:

✔ Dividing by 10 just requires moving a decimal point one place to the left. In that way, 27.5 (for example) becomes 2.75. If the number doesn't show a decimal point (7 for example), it really is there but not notated. Think of 7 as 7.0 — now you see a decimal point. When you move it one place to the left, the answer is 0.7.

✔ To divide by 5, first divide by 10 and then double the result. To divide 27.5 by 5, for example, first divide it by 10 (27.5 ÷ 10 = 2.75) and then double the result (2.75 × 2 = 5.50).

All looks confusing at first, but practice really helps.

Estimating with Ease

Estimating is the process of finding an approximate amount. The amount might be weight, volume, distance, time, or money. The result may not be perfect, but not it's expected to be. An estimation, although not precise, is good enough to get the job done.

When you take your car to the mechanic, you'll typically get an estimate of charges — the approximate (not the exact) sum of the costs of parts and labor. You also get an estimate of charges when you ask a crafts person to do painting, drywall work, or a fence installation. When you estimate, you're in good company. Corporations, economists, and governments make estimates all the time.

Apply the following simple techniques in estimating:

✔ **When possible, compare something known to what is unknown.** Known items could be parts of your body (for linear measurement), size of a bottle (for liquids), size of a box (for solid products), and so forth.

Use your body as a quick length estimator. The distance across the palm is about 4 inches (about 10 centimeters). The distance from the nose to the tip of the finger with the arm outstretched is about 36 inches (90+ centimeters).

✔ **Use a trick.** For example, in the United States, convenience stores have height scales painted on the door frames. It's considered a method of estimating the height of a fleeing armed robber, but it works equally well in giving you a quick estimate of your teenager's height.

✔ **Round up.** While you're shopping for groceries, round the prices up to the next 10 cents or dollar. As you add up your purchases, you'll get a good idea of what your bill will be. On a quick trip to the grocery store, you might think, "Let's see. A head of lettuce is about $1.40 and a soft drink is about $1.25. That's $2.65. I have $3.00 in my pocket. I can buy these things."

✔ **If all else fails, make a guess.** If your guess is based on any kind of reasoning, you may come close.

✔ **Do not perspire over minor details.** "Don't sweat the small stuff." An estimate is an approximation.

In the following sections, I cover some scenarios when estimating comes is really handy.

The rule of thumb for estimating is a practical one: If you can measure and calculate, then measure and calculate. If you can't measure and calculate, then estimate.

Estimating sales tax and value added tax (VAT)

Many states in the U. S. charge sales tax, and those states may have local variants that increase the tax. A safe estimate is to allow 10 percent of an item's purchase price for sales tax.

For example, if you buy a $70.00 item at the hardware store, you can determine what 10 percent of that amount would be by moving the decimal one place to the left (see the earlier section "Dividing in your head"). The tax will be approximately $7.00.

Elsewhere around the world, people pay a *value added tax* (VAT), which hovers around 20 percent, being higher in some countries and lower in others.

To estimate VAT, follow the same approach you use to estimate sales tax. Determine 10 percent by moving the decimal one place to the left in an item's price and then double the result; that gives you 20 percent. If an item costs €70.00, the VAT will be approximately €14.00.

Estimating tips

Diners in the United States typically tip about 15 percent of the sum on the check. Calculating a 15 percent tip is quick and easy, using the strategies I discuss earlier in this chapter. If, for example, the check is $34.57, drop the 57 cents to get $34. Determine what 10 percent of $34 is by moving the decimal one place to the left ($3.40). Divide that number by two to get the 5 percent ($1.70) and then add the two results ($3.40 and $1.70) together. Your 15 percent tip? About $5.00.

Estimating guests at a banquet

Do you want to estimate how many guests are attending a sit-down dinner at a wedding? Say you sit down at your table and count 10 seats available. (Banquet tables are usually round tables called *rounds,* and the guests are seated at "round of 8" or "round of 10" tables.) Stand up for a moment. Count the tables in the room. If you see 20 tables, a little mental multiplication tells you that the banquet has been set up for 200 guests (20 "round of 10" tables). Chances are that most seats will be filled, so the wedding has about 200 guests.

Doing Simple Cerebral Statistics

Professional statisticians spend a lot of time working with complex mathematics. However, most people aren't statisticians, yet they might still like to develop a statistic or two. The two handy stats you can do in your head are the average and the median.

Figuring averages

An *average* is the result of adding up numbers in a collection and then dividing that result by the number of items in the collection. The average is officially known as the *arithmetic mean.* Two averages are fairly easy to calculate using mental math — bowling average and grade point average.

Bowling average

Your bowling average is the total number of points you've bowled in a few games, divided by the number of games you've bowled.

Imagine that you bowl once a week in a league that meets 12 times. If you know your multiplication table up to the 12s and if your league bowls the typical three games in one night, you can see that you bowl a total of 36 games ($12 \times 3 = 36$). To get your average score for all those games, you add up all your scores and divide the total by 36.

Frankly, that's a lot of adding, and life is short. Why not just average your scores for the number of games you bowled in one night? (Besides, in many leagues, handicaps are calculated based on the average of the first night's bowling.)

Say, for example, that you bowl 150, 175, and 133 on the first night of the league. To find your average without lifting a pencil, do the following:

1. **Add the scores.**

   Break the numbers into more easily added combos, an addition trick I share in the earlier section "Adding numbers quickly." Here, for example, you "take" 25 from 175 and add it to 133, giving you 158. That reduces 175 to 150. Add 150 and 150 to get 300, and then just add 300 to 158 to get 458.

   $$\text{total} = 150 + 175 + 133$$
   $$\text{total} = 150 + (175 - 25) + (133 + 25)$$
   $$\text{total} = 150 + (150) + (158)$$
   $$\text{total} = 300 + 158$$
   $$\text{total} = 458$$

   The total score is 458 points.

2. **Divide 458 by the number of games, which, in this example, is 3 games.**

   You notice that 450 is a nice round number, so you decide to break the fraction into two fractions, which lets you divide 450 by 3 and then separately divide 8 by 3.

   Dividing 450 by 3 is easy. That division results in 150. The clumsy part is dividing 8 by 3, because the answer is "2 and something." Drop the "and something." Bowling averages are typically rounded down.

   $$\text{average} = \frac{458}{3}$$
   $$\text{average} = \frac{450}{3} + \frac{8}{3}$$
   $$\text{average} = 150 + 2\frac{2}{3}$$
   $$\text{average} = 152$$

Your average for the night is 152 points.

Grade point average

Many countries determine an average score for grades, but the systems vary. In the U. S., the grade point average (GPA) is a regular issue for students. Each letter grade (A, B, C, and D) is assigned a numerical equivalent (4, 3, 2, and 1, respectively).

To calculate a GPA in your head, do the following:

1. **Convert the grades to numbers.**

   Say you're taking 4 courses and your grades are A, A, C, and D. The numeric equivalents are 4, 4, 2, and 1, respectively.

2. **Add the numbers up.**

   This task is easy because most people don't take a lot of courses in one semester or quarter. The result is 11.

3. **Now divide by the number of courses — 4 in this example.**

   When you divide 11 by 4, you get a number that is greater than 2 and less than 3. Actually, it's 2 with a remainder of 3. Divide 2 by 3 to get 0.75. Your GPA is 2.75, or about a C+.

   $$GPA = \frac{4+4+2+1}{4}$$
   $$GPA = \frac{11}{4}$$
   $$GPA = \frac{8}{4} + \frac{3}{4}$$
   $$GPA = 2 + 0.75$$
   $$GPA = 2.75$$

Beware of the fallacy of averages (see the statistics section in Chapter 3). Most averages work best with a large number of items in a sample. Averages aren't as reliable with a small sample. For example, you could say, "I have one child who's four feet tall and one who's six feet tall. Their average height is five feet." That's mathematically correct, but it's meaningless.

Managing medians

The median is the value that separates the upper half of the items in a sample from the lower half. It's not the same as the average.

You can find the median by looking at the members of a group, arranging the values in order, and then counting down through the list until you reach the middle number. If the list has an even number of values, you take the average of the two central numbers.

Here's an example. Say you want the median age for your kid's Little League team, which has 11 players. You arrange the ages from least to greatest and count down to the middle, the 6th value in the list. That's your median. If your child's team has 12 payers, the median age would an average of the 6th and 7th values.

Often, you need to interpret, not create, a median. Imagine a neighborhood has 11 households. In 10 households, the annual income is $10,000; the 11th household has 1 millionaire with an annual income of $1,000,000. The average income is $100,000, but that doesn't mean that everyone's rich. The median income is $10,000.

You can do both of the calculations in your head by just looking at a list and counting the items. Simply count up from the lowest-value item to find the median.

Part II
Math for Everyday Activities

"We all know it's a pie, Helen. There's no need to pipe the number 3.14 on the top."

In this part . . .

*I*n this part, you go on a tour, but unfortunately not to any exotic places. You apply real-life math to real-life places — your home and stores around town.

With the math tips and tricks I share in these chapters, you'll be able to make smart choices at the grocery store and when shopping for other items. You'll be able to whip up things in the kitchen (even when you have to adjust a recipe), improve your nutrition and health, and take care of yard projects and home maintenance. The last chapter in this part takes you all over town and features a full workup of a modern problem: "Should I drive across town to save 10 cents per gallon on gasoline?"

Chapter 5

Let's Make a Deal! Math You Use When Shopping

● ●

In This Chapter

▶ Figuring out the actual cost of the items you buy

▶ Evaluating how much coupons, discounts, and sales really save you

▶ Looking at the costs associated with different payment methods

▶ Becoming aware of the impacts your shopping choices have

● ●

The practice of going to a marketplace is very old, so when you shop you're in the same company as citizens of ancient civilizations. In fact, historians think that Trajan's Market in Rome is the world's oldest shopping mall. You can bet that those shoppers wanted the same thing you do: to get what they wanted for the best price. Of course, shopping isn't always about bargains; it's about making satisfying choices, too.

The math you use when shopping is the same whether you live lean or high on the hog. It helps you make your choices. Although much of the math in this chapter takes place in the grocery store, the principles apply to shopping in all places. (And if a shopping trip just isn't a shopping trip unless you stop for lunch, too, flip to Chapter 9 for the math you use when dining out.)

Determining Actual Cost

Determining the cost of a purchase involves more than just looking at the price. The true cost is a combination of factors. Fortunately, the math is easy: Just figure out the various costs and add 'em up!

Finding the total cost of acquisition

The *total cost of acquisition* (TCA) is defined as all the costs associated with buying something. The TCA encompasses more than just the price; it includes taxes, delivery charges, and installation. Also included are the transportation costs — the money you spend to go get the item or to have it delivered.

Think of that washer/dryer set you've had your eye on. It's on sale, which is great, but you won't just pay the sale price. Taxes, delivery, installation, and the cost of new gas fittings (if you need them) will send the final, bottom-line price up. The common result? Sticker shock, where the total price you pay is a little (or a lot) higher than the advertised price.

Chances are excellent that you are already figuring TCA for purchases. For little things, you can do the calculations in your head. For big items, such as a car, you want to use a pad and pencil, a calculator, and/or a spreadsheet application. Following are some common scenarios.

Chasing a sale — Is it worth it?

Suppose you can buy a TV locally for $200.00, but you see that another store 32 miles away has it on sale for $190.00. As it turns out, your car gets 16 miles per gallon, so the trip will take 4 gallons of gasoline. If gas is $4.00 per gallon, is going to the second store worth it? To determine that, you add up the cost of the TV and the cost of the gas you'll use to get there (see Figure 5-1).

Item	Store 1	Store 2
Television	$200.00	$190.00
Gasoline (4 gallons)	$0.00	$16.00
Total cost of acquisition	$200.00	$206.00

Illustration by Wiley, Composition Services Graphics

Figure 5-1: Calculating whether a sale price far away is a better deal than a non-sale price closer to home.

As you can see, the answer is a big "No!" If you add in the cost of gasoline, you spend more money chasing the "bargain" than you save.

Comparing taxed versus non-taxed items

Some people avoid state sales tax by crossing a state line to shop in a state that doesn't have the tax. Buying over the Internet has the same effect. In this example, Store 2, where the item costs less but is 32 miles away, is in a no-sales-tax state. Add up the costs, and you can see that avoiding the sales tax means you save more by going to Store 2 (see Figure 5-2).

Item	Store 1	Store 2
Television	$200.00	$190.00
Tax (8 percent)	$16.00	$0.00
Gasoline (4 gallons)	$0.00	$16.00
Total cost of acquisition	$216.00	$206.00

Illustration by Wiley, Composition Services Graphics

Figure 5-2: Determining the better deal when sales tax is figured in.

Say that both stores charge the same sales tax rate (8 percent, for example). That means that the only factor affecting savings is the amount of gasoline you'll use.

To improve the savings, you can "spread" the gasoline cost over several items. In the example shown in Figure 5-3, you determine that, as long as you're going to get another television, you may as well buy a DVD player to go along with it, and (what the heck) how about buying that gas grill you like? If all these items are less expensive at Store 2, you can enjoy some serious savings by buying several items in one trip, since the same gasoline cost spreads over several purchases.

Item	Store 1	Store 2
Television	$200.00	$190.00
Gas grill	$300.00	$260.00
DVD player	$100.00	$85.00
SUBTOTAL	$600.00	$535.00
Tax (8 percent)	$48.00	$42.80
Gasoline (4 gallons)	$0.00	$16.00
Total cost of acquisition	$648.00	$593.80

Illustration by Wiley, Composition Services Graphics

Figure 5-3: Spreading out the cost of gas over multiple items.

This scenario isn't rare. You see this sort of buying every day at big box stores (also known as superstores or megastores).

Figuring the total cost of ownership

The *total cost of ownership* (TCO) is the estimated sum of direct and indirect costs of buying something. Direct costs are pretty much what you pay for an item; indirect costs come up later.

Indirect costs tend to occur again and again and again. The example in Figure 5-4 shows the real annual TCO for an entry-level automobile. Notice how much the total cost changes when you figure in both the direct and indirect costs.

Item	Monthly Direct Cost	Monthly Direct and Indirect Costs
Auto loan, $15,000.00		
48 month term/10 percent interest	$380.44	$380.44
Registration ($150.00 per year)		$12.50
Insurance		$105.00
Maintenance ($60.00 every other month)		$30.00
Gasoline		$173.20
Total cost of ownership	**$380.44**	**$701.14**

Illustration by Wiley, Composition Services Graphics

Figure 5-4: Factoring in both direct and indirect costs.

The math is easy: Just figure out the costs and add 'em up! The car payment isn't that bad (only $380.44), but the TCO ($701.14) may cause you some pain. (And this doesn't even include depreciation — the concept that the selling price or trade-in price of a car goes down a little every day.)

There are two valuable lessons here. First, with TCO analysis, you see the true cost of continued ownership, and second, such exercises get you thinking about costs you may have ignored.

Calculating TCO is useful for what economists call *durable goods* (long-lasting items such as cars, furniture, appliances, and home electronics). You'd never try to calculate TCO for a vacation, because doing so just doesn't make any sense. With a vacation, the money is spent; you have the memories and the photographs, but no long-term cost of ownership. Accountants call spent money *sunk costs*. The best cost analysis for a vacation consists of planning for every expense and comparing prices for major expenses such as airfare and hotels. Jump over to Chapter 9 to see how it's done.

Uncovering hidden costs

Hidden costs are usually expenses not included in the purchase price of an item. Mostly, nobody's trying to hide them; they are simply the costs of supplies, installation, maintenance, or minor recurring fees. However, there are other costs that merchants would prefer you didn't know about. These are expenses that aren't prominent in sales literature. The old saying that "the big type giveth and the small type taketh away" applies. Those costs come to light later, often at a point where you feel it's too late to back out of the deal.

Places where hidden costs are common include the following:

- ✔ **Cruises:** Cruise lines advertise their rates as "all inclusive," which is true as far as your stateroom, meals, and some activities go. But you pay extra for onshore excursions, massages, an Internet connection, and cocktails.

- ✔ **Airfare:** Until recently, the full price of an airline ticket in the United States wasn't revealed until the purchasing process was almost complete. That is, the advertised price was lower than the real price. Various taxes and fees weren't included, and some still aren't. You can expect many other new, strange fees, too, including a fee if you want priority boarding and charges to check in luggage. Figure 5-5 shows how a $378.00 fare is really a $420.80 fare. The excise tax is part of the advertised fare, but all the other fees are still separate.

- ✔ **Bank charges:** Banks in the United States have long invited controversy about fees for monthly service, overdrafts, and overdraft protection. The trend in legislation is to demand greater transparency.

- ✔ **Unbundling:** *Unbundling* is separating the price of goods or services from a single charge into separate charges. Every buyer expects something to be "not included" with the purchase, but unbundling is a merchant's deliberate attempt to show lower prices by separating (that is, not including) some fees from the basic cost of the product. For example, you can see unbundling when you buy smartphones, e-readers, and tablet computers. The basic "box" is nothing other than the device, and you have to buy screen protectors, cases, and so forth, separately.

Cost and Payment Summary	
Base Fare	$351.62
+ Excise Taxes	$26.38
Advertised Fare	**$378.00**
+ Segment Fee	$14.80
+ Passenger Facility Fee	$18.00
+ Security Fee	$10.00
Total Payment:	**$420.80**

Illustration by Wiley, Composition Services Graphics

Figure 5-5: Airfare fee summary: Notice the additional charges.

Hidden cost math is very simple. Locate the hidden costs, if you can, and add them up.

total cost = advertised cost + hidden costs

In the airline ticket example, this is the calculation of hidden costs:

total cost = advertised cost + hidden costs

total cost = (base fare + excise tax) +

(segment fee + facility fee + security fee)

total cost = (\$351.62 + \$26.38) + (\$14.80 + \$18.00 + \$10.00)

total cost = (\$378.00) + (\$42.80)

total cost = \$420.80

Making Tradeoffs: A Fun Balancing Act

Consumers in many countries have abundant choice in what merchandise they buy. Some common choices include

- ✔ A choice between a name brand product and the generic product (also called a store brand or private label product).

- ✔ A choice between organic and non-organic products or locally grown produce and produce from another country.

- ✔ A choice between a product made in your own country and one made on foreign shores.

When faced with many choices, consider the tradeoffs. A *tradeoff* is a buying scenario where you decide to give up one thing in order to get another. Typical tradeoffs might be

- ✔ **Price versus name brand:** When you're buying food, you may opt for a generic product because you think the name brand costs too much. Conversely, you may go for a more expensive, name-brand television because you have confidence in that manufacturer's products.

- ✔ **Quality versus price:** You may decide to buy the more expensive suit with a well-known designer label because you're confident it's well made and will last a long time.

- ✔ **Other factors versus price:** You think, "The product made in my country costs more, but I feel patriotic by supporting it. I'll buy it instead of the foreign product." Or you think, "The organic produce costs more, but I think it's healthier, and it's locally grown. I'll buy it, despite its higher price."

Price is almost always a factor in making tradeoffs because it's easy to quantify the money. If price is the only factor in a tradeoff, the lower price almost always wins. The subjective factors in a tradeoff — like the value you place on organically grown food or your feelings about buying products made in your own country — are important, too, even though they're very difficult to quantify.

Don't be afraid to quantify subjective factors. One way to do it is to assign a number from –10 to +10 for each consideration, with –1 through –10 representing the downside of making the purchase, and +1 through +10 representing the upside. For example, say it's your 20th wedding anniversary, and you're thinking of getting your ever-lovin' honey a nice gift. Figure 5-6 shows what your reasoning might look like, with numeric ratings. In this instance, when you add up the ratings, the gift gets a +2. Buy it.

Item	Rating
This gift will cost $200.00	–3
But it's our 20th anniversary	6
But I only make $200.00 per week	–5
But my spouse will be impressed	3
And I can charge it	1
But I have to pay off the charge	–1
What the heck, you don't live forever	1
Tradeoff value	**2**

Illustration by Wiley, Composition Services Graphics

Figure 5-6: Assigning numeric values to the intangibles.

Where to buy is another, but different, kind of tradeoff. See the section "Determining Where to Shop."

Buying in Quantity: A Good Deal?

Buying in quantity allows you to take advantage of *quantity pricing*. With quantity pricing, a merchant offers a lower price if you buy more of a product. Many people love such pricing and like to call it a discount (which it isn't).

Buying a quantity of a product costs more, but the unit price is lower. To figure the unit price, you divide the total price by the number of units:

$$\text{unit price} = \frac{\text{total price}}{\text{number of units}}$$

Compare two extremes of unit pricing. A single bottle of water at a convenience store might sell for $1.00, making its unit price $1.00:

$$\text{unit price} = \frac{\$1.00}{1}$$

unit price = $1.00

By contrast, you can buy a case of bottled water at a giant warehouse store for $4.45. If the case contains 35 bottles of water, its unit price is about $0.13 per bottle — quite a price difference compared to the convenience store bottle!

$$\text{unit price} = \frac{\$4.45}{35}$$

unit price = $0.127

To calculate unit prices, use a calculator. Over time, you may find that you can make good approximations simply by doing the math in your head.

 Quantity pricing is super when you use a lot of an item, such as bottled water or paper towels. Common sense says that it's not a good idea to buy big quantities of items you using sparingly (or rarely).

Knowing the Real Cost of Sale Items

A *sale* is a temporary reduction in the price of an item. Many merchants, big or small, local or national, have sales. Sales are popular because the idea of getting an item at a lower price than previously advertised appeals to many people.

Sales may be seasonal (think post-Christmas), or pseudo-seasonal. A *pseudo-seasonal sale* is a "manufactured" sale, often citing a holiday. For example, think "Big Fourth of July Blowout!" No real reason exists for a sale on the Fourth of July. A *closeout sale* or *clearance sale* is intended to reduce the inventory of an item to zero.

In your quest for a deal, be aware that not all sales — or bargains or discounts or coupons — are created equal. In the following sections, I tell you how to decipher which deals are worth pursuing.

Bargain buying rules and cautions

Judging a good deal is entirely subjective. For example, I have a friend who bought a used Rolls-Royce, and he felt that he got a very good deal. I don't need a Rolls-Royce, so the deal wouldn't appear to me to be a bargain.

So before you plop down your hard-earned cash, ask yourself these simple questions when you find a "bargain." Doing so will help you avoid making an irrational — and sometimes costly — decision:

✔ **Do I need the item?** If you need the item, it doesn't matter if it's on sale, but so much the better if it is. If you were hesitating about buying this item, maybe a bargain price will get you to act.

✔ **Is it "too good to pass up?"** Really? If you don't need the item, the benefit doesn't exceed the cost. Don't buy.

✔ **Can I afford the item?** Not being able to pay for an item pretty much reduces the joy of owning it.

Be prepared for "bargain hunter's remorse." Later, you may realize that you could have done without the item entirely, or you may find that you could have bought it for less somewhere else.

Counting coupons

Some sales apply to everyone. Other sales are based on a coupon and apply only to people who present the coupon to get a reduction in price. (Interesting tidbit: The word *coupon* comes from the French *couper*, meaning "to cut," and you generally cut coupons out of the newspaper.)

Do coupons really save you money, and if so, how much? The following sections help you find out.

Single item coupons

The most common coupon is for a fixed amount off a particular item. Think "10 cents off on a 15 ounce can of Aunt's Tillie's Baked Beans," for example.

What does $0.10 off really mean? It's an absolute reduction of $0.10. It doesn't amount to a hill of beans whether Aunt Tillie's Baked Beans normally sells for $1.49 or $1.99; the coupon still reduces the price by 10 cents.

Grocery coupons almost always apply to brand name items. Compare the sale price of the brand name item to the price of its generic equivalent to see where the bigger savings are.

All-item coupons

You may receive a coupon that applies to all or almost all items in a store. Figure 5-7 shows such a coupon.

Unfortunately, the coupon requires a $75.00 minimum purchase. You have to spend $75.00 to save $10.00. If you don't need $75.00 worth of merchandise, the coupon isn't useful.

Illustration by Wiley, Composition Services Graphics

Figure 5-7: An all-item coupon.

Calculating percentage decreases: You save 10 percent!

Sometimes you see an advertised discount, such as "Take 10 percent off any item in the store." If the deal applies to everything in the store, that's good. If it applies to the total cost of multiple items, that's better yet!

A 10 percent discount is easy to calculate. In fact, you can do it in your head. Just take the item's price and move the decimal point one place to the left. A 10 percent discount on a $70.00 item is $7.00. Alternatively, you can perform this calculation:

discount = regular price × 0.10

discount = $70.00 × 0.10

discount = $7.00

Figuring the discounted amount is easy, too. Just subtract the discount from the regular price.

discounted price = regular price − discount

discounted price = $70.00 − $7.00

discounted price = $63.00

Calculating the real percentages in "get one free" offers

"Get one free" offers come up from time to time. The most common is "Buy one, get one free," but occasionally you see other versions (like around July 4th, when fireworks retailers offer "Buy five, get one free" deals). Calculating the average price is as simple as dividing the price for one item by the number of items. In the "buy one, get one free" scenario, you divide by 2 items.

$$\text{average price of item} = \frac{\text{regular price} + \text{price of free item}}{\text{number of items}}$$

$$\text{average price of item} = \frac{\$7.00 + \$0.00}{2}$$

$$\text{average price of item} = \frac{\$7.00}{2}$$

average price of item = $3.50

The average price is $3.50 per item. When a $7.00 item sells for $3.50, that's a 50 percent discount.

"Buy two, get one free" offers are also common. The math is similar.

$$\text{average price of item} = \frac{\text{regular price} + \text{regular price} + \text{price of free item}}{\text{number of items}}$$

$$\text{average price of item} = \frac{\$7.00 + \$7.00 + \$0.00}{3}$$

$$\text{average price of item} = \frac{\$14.00}{3}$$

average price of item = $4.67

The average price is $4.67 per item. When a $7.00 item sells for $4.67, it's selling for 67 percent of its regular price. That's a 33 percent discount.

A typical variation is what I call the "Buy 1 and Don't Get Another One Free." These coupons will offer a second item but not for free. Usually a small cost is associated with it, as well as a requirement to buy some other item. The coupon in Figure 5-8 gives you a second sandwich for $1.00, as long as you buy the first sandwich and a beverage.

Hoagie Haven

$1 TURKEY & BACON AVOCADO 6" HOAGIE!

When you buy a 6" Turkey & Bacon Avocado hoagie and a drink.

Offer only valid at participating Hoagie Haven locations. This coupon cannot be used in conjunction with any other offers. Cannot be redeemed for cash. Additional charges may apply. Not for sale. All rights reserved.

Expires 8/17/12

Illustration by Wiley, Composition Services Graphics

Figure 5-8: A variation of the typical "buy one, get one free" deal.

Dealing with dining specials

Restaurants have a coupon variation that you've probably seen. "Buy one entrée at full price and get a second entrée (of equal or lesser value) for half price." Is that a deal, or what?

You develop the answer with a tried-and-true costing technique: List the factors, figure in the tax and tip, and add everything up. Figure 5-9 shows you how.

Item	Without Coupon	With Coupon
Entry #1	$15.00	$15.00
Entry #2	$15.00	$7.50
Beverages	$4.00	$4.00
SUBTOTAL	$34.00	$26.50
Tax (7.75 percent)	$2.64	$2.05
SUBTOTAL	$36.64	$28.55
Tip (15 percent)	$5.50	$4.28
TOTAL	$42.14	$32.83

Illustration by Wiley, Composition Services Graphics

Figure 5-9: Calculating savings on a "buy one, get one half off" deal.

Entrée #1 is $15.00 with or without the coupon, but with the coupon the price of Entrée #2 drops from $15.00 to $7.50. That's a nice savings. Beverages aren't part of the deal.

In theory, you should tip on the undiscounted prices. In real life, the server takes his or her chances. On the other hand, you shouldn't tip on taxes, but people do anyway.

Doubling down on discounts

Discounts are consistent reductions in the basic price of goods or services. That's a bit different from coupons and sales. While coupons expire and a sale may last a day or a week, discounts tend to apply all the time.

The merchant discounts prices in the hope of attracting customers who might not otherwise be able to buy (seniors on a fixed income, for example). Also, the merchant may want to increase traffic during otherwise slow times (for example, matinee showings at movie houses and pre-dinner hours at coffee shops).

The purpose of discounts is to attract and build business. For that reason, merchants can be infinitely resourceful about offering them. You can find an incentive discount for practically everyone. You see student discounts, employee discounts, military discounts, child discounts (think "Kids eat for free!"), and senior discounts. Other discounts are available if you're a member of an organization. A prominent example is the American Association of Retired Persons (AARP), which offers members discounts on practically everything, including travel, entertainment, prescription drugs, and insurance.

Your real-life math task is to calculate discounts and make good math-based judgments.

Basic discount math

Discounts can come in various forms. Some discounts are a percentage ("10 percent off to seniors"), and some are a lower price ("$2.00 off for seniors"). Some discounts are in effect for part of the day ("early bird" specials at the coffee shop, for example).

Most discount math is incredibly simple. Just subtract the discount. For example, if a regular ticket price is $8.00 but seniors or students get $3.00 off, the cost of the ticket is $5.00.

No percentage calculations are required for this coffee shop special: "Two early bird dinners for $12.00." What could be simpler than dividing by 2?

$$\text{discounted price per meal} = \frac{\text{total price}}{\text{number of diners}}$$

$$\text{discounted price per meal} = \frac{\$12.00}{2}$$

$$\text{discounted price per meal} = \$6.00$$

To see your savings, just subtract the discounted price from the regular price. If a meal is normally $8.95, and you're paying $6.00, you save $2.95.

Determining the real discount in "double discounts"

You'll love this. The ad says "Double discount! Everything in the store 25 percent off! And take another 25 percent off at the register!"

You get a 50 percent discount, right? Wrong, wrong, wrong! It's really a 44 percent discount. Look at this $10.00 item when double discounted:

$$\text{final price} = \text{regular price} \times 0.75 \times 0.75$$

$$\text{final price} = \$10.00 \times 0.75 \times 0.75$$

$$\text{final price} = \$5.625$$

$$\text{final price} = \$5.63 \text{ rounded}$$

A discount of 25 percent means that you're paying 75 percent of an item's regular price. When you calculate the discount two times, the answer is *not* 50 percent. The second discount applies to the first discount amount. In the example, the first discount takes the purchase price from $10.00 down to $7.50. Then the second discount is applied, taking the final price to $5.63. That's not the same thing as $5.00, the amount you probably expected to pay.

How Do You Wanna Pay for That?

When you shop, you can pay in several ways: cash, check, money order, on account, debit card, PayPal, and credit card. Each method has advantages and disadvantages.

✔ **Paying with cash:** This form of payment completes a transaction. There are no debit card entries or credit card statements to worry about. You rely on the merchant to stand behind the purchase.

✔ **Paying by check:** Checks represent cash, but some merchants are reluctant to take them. They feel that they've been burned too many times by bad checks.

✔ **Paying with a money order:** Money orders are a useful way to pay some bills by mail. Because you buy them with cash, you have no debt worries.

✔ **Buying on account:** You can buy "on account" at some stores, a method often used by small businesses. You get a monthly statement and then write one check to the merchant for all purchases made during the month.

✔ **Paying with a debit card:** Also known as an ATM card, purchases with a debit card complete a transaction. As soon as you pay, the amount is almost instantly removed from your bank account.

✔ **Paying with PayPal:** PayPal is usually required for eBay purchases and is often an option with other online vendors. Paying with PayPal is a little like paying with a debit card, because PayPal immediately debits your checking account.

✔ **Paying with a credit card:** Credit cards are convenient and honored almost everywhere. The merchant is paid by the credit card issuing bank, and you pay the issuer when you get a monthly statement. Because it's essentially a consumer loan, you decide how much and when to pay it off. Credit cards offer some protection if a merchant won't make good on a defective product, because the bank that issues the credit card will usually reverse a charge if you have a dispute.

Capturing bargains with credit cards

Credit cards are one of the most convenient ways to pay. In fact, sometimes they are too convenient, because they can put you in serious debt (which I explain in Chapter 10). But they do offer a little-known advantage: The credit card is a great tool for capturing bargains. For example, if a giant big screen HDTV is on sale for $750.00 and it's normally $1,500.00, maybe you should buy it now. Most likely, you will use a credit card to pay for it.

Another advantage to paying by credit card is that you can choose how fast you want to pay down your credit card balance. But be careful, because that advantage can turn into a liability, and you may be paying off that HDTV forever.

A disadvantage is that you have to pay interest, typically at 18–21 percent per year, if you don't pay off your balance each month. Another disadvantage is that you may go overboard and buy more than you can realistically pay off.

Follow these basic rules to use credit cards wisely:

- ✔ **Ask yourself whether the purchase is worth it.** If the item is essential (for example, school clothes for the kids), use the credit card.

- ✔ **Ask yourself when the item will wear out.** If the item is used up before it's paid off, buying with a credit card is a bad idea. This is especially true of meals and admission to theme parks. If the purchase will last a while (like that great power saw at Home Paradise), it's probably okay to charge it.

- ✔ **Ask yourself whether paying by credit is more convenient.** Paying with a credit card is far more convenient when you're not sure exactly how much cash you'll need to make the purchase (an expensive meal out, for example).

- ✔ **Ask yourself whether you can pay off the charge.** Remember, that a little bit of every purchase contributes to your credit card balance, until the balance is zero (see Figure 5-10). The only way to avoid this "growing balance" is to pay off your balance each month.

In this example, you've made five payments of $50.00. That's $250.00. But if the interest rate is 18 percent per year, you've only lowered your balance by $191.93. Note that you charged both an HDTV and a dinner. The HDTV will last a long time, but that dinner lives only on your waistline.

New Charges	Starting Balance	Interest	Monthly Payment	Ending Balance
HDTV $750.00				
Dinner $100.00	$850.00	$12.75	$50.00	$812.75
	$812.75	$12.19	$50.00	$774.94
	$774.94	$11.62	$50.00	$736.56
	$736.56	$11.05	$50.00	$697.61
	$697.61	$10.46	$50.00	$658.07

Illustration by Wiley, Composition Services Graphics

Figure 5-10: A growing credit card balance.

You can use your math skills to build a payoff chart. Take the starting balance, add in interest (at the monthly rate of 1/12 the annual percentage rate), and subtract the payment. You will get an ending balance.

For advice and strategies for avoiding or getting out of credit card trouble, head to Chapter 10.

Taking advantage of layaway

Layaway is a way to buy an item without paying for it all at one time but avoiding using credit. The merchant "lays away" the item for you in storage, and you don't get it until you've completely paid for it.

Here are the advantages of layaway:

- ✔ After you've paid for it completely, the item is yours, all yours. If you change your mind and don't complete the transaction, the item goes back into stock and you get your money back (but you're charged a small fee).

- ✔ The price is fixed, and you're not charged interest (as you would be with a credit card).

- ✔ The availability of the item is guaranteed, because it has been laid away.

- ✔ As a bonus, you may also gain a sense that you're living within your means.

Layaway was a big deal during the Great Depression, and it has been making a comeback, especially during the last several Christmas shopping seasons. Why Christmas? Because that's the season when many retailers make most of their money, and they want to encourage sales.

Discovering Deals at the Grocery

The supermarket is a super place to compare prices and discover deals. Why? The Food Marketing Institute reports that, as of 2010, the average grocery store has about 38,718 items. Giant stores have as many as 60,000 items!

In this section, I share the math you need to know to estimate your grocery bill, estimate how much you need to buy, and find the best deals.

There are several simple estimates you can make. You may be making them already when you shop. The whole point is to find the greatest benefit with the least cost.

Estimating the whole grocery bill

You can estimate your entire grocery bill by keeping a running total in your head, on a piece of paper, or on a smartphone. To make the task simpler, just round up each item's price. For example, if a can of vegetables is priced at $0.89, call it $1.00. When you reach the checkout, your estimate will be a little high, and so you won't be surprised at the total.

If it seems like too much effort to estimate the sum of all items, try doing it with only the most expensive items you buy. That estimate will give you a sense of how "big impact" items are affecting your grocery bill.

In trying to determine your whole grocery bill, don't forget to include taxes. For most stores, just add 10 percent. Take your estimated balance and move the decimal point one place to the left. For a purchase of $70.00, estimate paying about $7.00 in sales tax. Because some states don't charge sales tax, other state and local sales taxes haven't hit 10 percent (yet), and many don't tax food items and medicine, your estimate will be high. Which means you'll get a pleasant surprise at the checkout.

Estimating how much to buy

Charity begins at home, and so does estimating. You benefit by knowing how much to buy before you turn yourself loose at the grocery store. For example, if you'd like to serve a standing rib roast at a big family meal, know your needs in advance:

- ✔ To feed 6 people, buy a 3-rib roast.
- ✔ To feed 8 people, buy a 4-rib roast.
- ✔ To feed 10 people, buy a 5-rib roast.
- ✔ To feed 12 people, buy a 6-rib roast.
- ✔ To feed 14 people, buy a 7-rib roast.

The math is simple. Begin by looking up recipes to determine what an average serving size is (the Internet is a good place to look), and then multiply the serving size by the number of guests. For a rib roast, for example, you'll want to allow 1 rib for every 2 guests. Be careful! The last time I looked, a full 7-rib roast cost about $104.00!

Steamed clams are easier. There are generally 12 to 15 clams per pound. Serve 1–2 pounds of steamer clams per person (most of the weight is waste — the shell). If you're serving 4 guests, you'd want 4–8 pounds.

Comparing unit prices

Grocery deals are based almost entirely on price. Generally, the lowest price wins. When you're comparing items, try to compare apples to apples. Shallots, for example, cost about six times what yellow onions do, so comparing shallots to yellow onions isn't really fair.

You calculate deals based on unit price. Sometimes, the store practically gives this information away; sometimes, you have to calculate to get it.

Some grocery stores feature unit pricing on the shelf, not on the item. Figure 5-11 is an example of unit pricing label.

Illustration by Wiley, Composition Services Graphics

Figure 5-11: A unit pricing label.

Look carefully. The cost of a jar of pickles is $3.99, but the unit price is $0.065 per ounce. If you know the unit price, you can easily compare this product to other brands and other sized jars of pickles.

Comparing products per roll or square foot

You can compare items such as aluminum foil, plastic wrap, paper towels, and toilet paper in several ways. Methods include per roll and per square foot.

Comparing prices of the same brand of paper towels by package size is easy. You can practically do it in your head. Just divide the price by the number of rolls in the package. Figure 5-12 shows a comparison between a single roll of paper towels, selling for $2.59 and a 3-roll package selling for $6.29.

Comparing Number of Rolls	Price	Price Per Roll
Paper towels – 1 roll	$2.59	$2.59
Paper towels – 3 rolls	$6.29	$2.10

Illustration by Wiley, Composition Services Graphics

Figure 5-12: Comparing one to many.

In Figure 5-12, you can see that the unit price for 3 rolls is less than the unit price for 1 roll.

For aluminum foil or plastic wrap (or anything sold by the square foot), you compare by square footage. The comparison is easy, because the number is right on the package. Figure 5-13 compares three rolls of aluminum foil.

Comparing Square Footage	Price
Brand name – 50 SF	$4.49
Store brand – 50 SF	$2.99
Store brand – 75 SF	$2.50

Illustration by Wiley, Composition Services Graphics

Figure 5-13: Comparing by square foot.

When you compare 50 square feet (SF) of the brand name foil with 50 square feet of the store brand, you see a big difference in price. But wait! Who would have expected that the store had 75 square foot rolls on sale? That 75 SF roll costs less than the 50 SF roll.

Calculating volume (25 percent more free!)

Liquid volumes are calculated in fluid ounces and in liters. Comparing for value can be incredibly simple. When quantities are the same (for example, 2-liter bottles of soft drinks), just compare the prices. The same is true when comparing six-packs of 12 ounce (355 mL) soft drinks.

Comparing two items with different volumes gets a little tricky. But after a while you can do it in your head. The idea is to make the different volumes more or less equivalent to each other. For example, a six-pack of cola has 6 cans with a capacity of 355 mL. Think "2 cans contain about 700 mL, so 6 cans contain about 2,100 mL." Well, 2,100 mL is only a little more than the volume of a 2-liter bottle (2,000 mL). Say the six-pack sells for $2.99, while the 2-liter bottle *of the same product* sells for $1.99. Buy the big bottle.

Determining Where to Shop

Your money is important, not just to you, but to retailers. Every time you shop, you're essentially voting for something — for a product or a merchant. If you think about it, that makes you a very powerful person. A store may not know you as an individual, but as part of the buying public, you can help its prosperity or put it out of business.

Narrowing your choices

A key component is choosing where to shop. Each type of store can fill a need, so where you shop depends on your needs. You might go to a superstore to save money on groceries, but you might also go to a specialty grocery store to find Italian, Jewish, Mexican, Chinese, or Thai items you can't get anywhere else. There are also specialty stores that sell only meat, fish, or cheese.

Narrow your shopping choices down to three types: local store, big box stores, and online stores. Each type of store has advantages and disadvantages for you:

✔ **Local stores:** These stores are close by. They are usually owned by people in your community. Some stores (think Kmart or Safeway) are "corporate" stores but employ people in your community.

✔ **Big box stores:** These stores may be near you, or they may require a bit of a drive. They are known for their giant size and very low prices.

✔ **Online stores:** Online stores have no "brick and mortar" store at all. They are highly convenient places to shop. However, you can't hold the merchandise in your hands. A nice delivery service will bring your purchase to you, so you don't have to drive a car (but you do have to pay for shipping). You don't even have to dress up, for that matter.

Looking at externalities

These days, every consumer has to be an economist. It seems as though you need to know everything from the cost of the money that funds your mortgage to the value of the *renminbi* (the currency of the People's Republic of China). So you should know about externalities.

An *externality* is a cost or benefit that isn't transmitted by prices. A *negative externality* is a cost that the creator of a problem doesn't bear. "Nobody" pays for it, except that we all pay for it. You could call it "playing for free." For example, if a chemical company pollutes a river, "nobody" pays for it, except that the costs of cleanup come out of everyone's taxes. If a local store closes, "nobody" except the store takes a loss, except that that community may have new, unexpected costs to bear.

Rest assured that shopping can produce negative externalities. For example, if you drive a long way to a superstore, the air gets polluted and the roads wear out. Negative externalities are difficult, but not impossible, to compute, and they are reported in newspapers and online. As consumers become more aware of them, they demand changes. It's mainly a matter of considering the overall costs and benefits of your shopping choices.

There are also *positive externalities*, too. For example, when a new store opens, the store hires employees. The employees buy homes and pay taxes that help the schools.

The impact of your shopping choices

Where you spend your money determines the success or failure of businesses. You can have a big impact on business, which is only fair, because some businesses can have a big impact on you.

The study of economic and social impacts is a broad field. You just need to know the basic principles. By knowing the principles, you can use your math skills to get a sense of the impact your decisions about where to shop have. If the concept can be boiled down to one general principle, it would be that everything affects everything. Your shopping choices matter!

Gauging the impacts of different store types

Each store type has both direct and indirect impacts on a community. Consider what happens when a big box store opens and local merchants go out of business. To put it simply, an *indirect impact* is when your neighbor loses his or her job. A *direct impact* is when *you* lose your job.

Impact of the big box store

The biggest impact of a big box store is that it may cause local business to fail. For example, if a local supermarket closes, you can calculate the cost of this closure by using simple multiplication. Just multiply the average salary by the number of employees out of a job. For example, if the average salary is $20,000.00 and the local store had 93 employees, your community has lost $1,860,000.00.

$$93 \times \$20,000.00 = \$1,860,000.00$$

In other words, the economic impact of the store closing is a loss of about $1.9 million.

The social impact can be far-reaching, too, especially if the laid-off employees can't easily get new jobs. That situation may stress the Temporary Assistance for Needy Families program (sometimes called welfare) and the resources of the churches and food banks.

When a local store stays open, there's also a "multiplier effect." Store employees spend their paychecks at other local stores and make donations to local charities.

The prices at big box stores are very low, but if they require a drive to get there, you create a bigger carbon footprint than if you had shopped locally. And, bluntly, some big box stores have a reputation for paying low wages and not offering healthcare benefits.

Impact of online shopping

Shopping online may or may not produce lower prices. Sometimes, the base price "giveth," but shipping costs "taketh away." Figure 5-14 shows a price comparison between buying a TV online and buying one at a local store. The math is a simple cost comparison. Find the items that make up the total cost of acquisition, and add everything up.

Item	Local	Online
Television	$200.00	$180.00
Delivery	$0.00	$24.00
Tax (8 percent)	$16.00	$0.00
Total cost	$216.00	$204.00

Illustration by Wiley, Composition Services Graphics

Figure 5-14: Comparing online and local shopping.

In this example, you avoid paying sales tax (although this is changing), but you pay for shipping. In the end, you save only a few dollars. When the government requires all online vendors to charge sales tax, you may end up paying *more* for an online purchase than a local purchase.

Online shopping reduces your carbon footprint. It also reduces the volume of business for local merchants. You also give up the opportunity to handle the merchandise.

Online purchases require you to use a credit card, debit card, or PayPal to make the purchase. If you feel uncomfortable using these payment methods online, you probably won't be able to buy online.

Be careful of online purchases where the item costs $5.00 but the shipping cost is $10.00. This really happens!

Chapter 6

Mmm, Mmm, Good: Kitchen Calculations

*T*he kitchen's a great place to hone your math skills. After all, you're doing math already, whether you know it or not. Having a few math tools and simple techniques at your disposal makes cooking better and easier, and the results of your math work will taste great!

In this chapter, I tell you how to convert from one measurement unit to another (a very common dilemma in kitchens everywhere), how to scale a recipe up or down so that you make just the number of servings you need, and how to apply math to some everyday cooking and purchasing tasks.

Taking Measure

Measuring ingredients correctly makes you a better cook (well, at least, a more accurate cook). To measure ingredients accurately, you need to understand the most common measurement units in the kitchen, and you need to know a bit about equivalencies — both of which are covered in the following sections.

Knowing your units

In the kitchen, you measure (and sometimes convert) volume and weight. In order to perform that task, you need to understand *systems of measurement* and *units of measure*.

The United States uses the U. S. customary system of measurement (sometimes called the *American system*). Some of the units are very common in the home, especially in the kitchen, and are often referred to collectively as the *household system*.

The American units you will likely encounter in the kitchen are

- **Mass:** Mass (weight) is measured in ounces (oz) and pounds (lb).

- **Liquid volume:** Liquid volume is measured in fluid ounces (fl oz), pints (pt), quarts (qt), and gallons (gal). Other common units include teaspoons (tsp), tablespoons (tbsp), and cups (C).

In the past, the *imperial system* (also known as the *British Imperial*) was the rule for countries in the British Empire. The system has (among other things) a larger gallon than the American system. Since 1995, the United Kingdom has used the metric system. So does every country in the world, except for the United States, Burma (Myanmar), and Liberia.

The metric units you will likely encounter in the kitchen are

- **Mass:** Mass (weight) is measured in grams (g) and kilograms (kg).

- **Liquid volume:** Liquid volume is measured in milliliters (mL), deciliters (dL), and liters (L).

- **Length:** Length is measured in millimeters (mm) and centimeters (cm).

You use a couple other measurements every day in the kitchen, but they don't come up on any formal list of units:

- Temperature, measured in degrees Fahrenheit or Celsius

- Time, measured in seconds, minutes, and hours

- "Per each" units or fractions of units, such as butter (that is, 1 stick) and fruit or vegetables (that is, 1/2 lemon, 2 medium onions, 2 eggs, and so on)

Temperature and time are pretty much things you just set (but don't forget), but you may find yourself needing to change "per each" units according to your needs.

Equivalencies — All things being equal

Which is more, 4 fluid ounces of milk or 1/2 cup of milk? Neither. They're both the same quantity. And that's what an equivalent is. How many teaspoons are in a tablespoon? That's another equivalent. The answer is 3.

Knowing things like how many pints are in a quart or how many teaspoons are in a tablespoon comes in mighty handy in the kitchen, especially when you're scaling recipes (the topic of the next section) or when you don't have the appropriate tool to measure in the unit the recipe specifies (say the recipe calls for fluid ounces but you need to know what it is in cups).

The solution? Memorize the equivalencies you use most or trust a handy-dandy source, like Table 6-1, which lists common equivalencies that practically do the work for you.

Table 6-1	Common Equivalencies	
Name	*American Equivalent*	*Metric Equivalent*
1 teaspoon		5 milliliters
1 tablespoon	3 teaspoons	15 milliliters
1 cup	8 fluid ounces	240 milliliters
1 pint	2 cups, 16 fluid ounces	480 milliliters
1 quart	2 pints, 4 cups, 32 fluid ounces	960 milliliters, 0.96 liters
1 gallon	4 quarts, 8 pints, 16 cups, 144 fluid ounces	3.785 liters
1 deciliter	3.38 fluid ounces	100 milliliters
1 liter	33.8 fluid ounces	1,000 milliliters
1 ounce		28.35 grams
1 pound	16 ounces	454 grams
1 inch		25.4 millimeters, 2.54 centimeters
10 inches (cucumber, for example)		25.4 centimeters

Most kitchens have the basic tools (measuring spoons and measuring cups) that make kitchen math possible. If you do a lot of cooking (and hence a lot of kitchen math), two great options are a digital scale and a digital timer. Neither is absolutely necessary, but both can help you. Without them, you may end up doing "time math" and "weight math" the whole time you're cooking.

- ✔ The digital timer counts down to show how much cooking time's left for a dish, which is much more convenient than looking at the clock, adding cooking time to determine the time when a dish will be done, and then having to watch that clock. The timer can also count up to show how long something has been cooking.

- ✔ The digital scale is an inexpensive, easy, precise way of measuring ingredients by weight in both American and metric units. It sure beats guesstimating, which nobody does very well. To me, 3 ounces, 4 ounces, and 5 ounces of an ingredient all look about the same.

Scaling a Recipe

Typical recipes are great if you are an average household, but most families aren't average. Many people, young and old, live alone, and many couples delay having children for a while. Young children eat like, er, children, but older children eat more like adults. Also, a recipe for a family won't work when friends (or all the relatives) are coming over. That means that most recipes need some adjustment.

So what do you do when a recipe says it will yield four or six servings, but you want to feed more or fewer people? You change the recipe to produce the number of servings you want. This is called *scaling* a recipe. To increase the number of servings, you scale up a recipe. To decrease the number of servings, you scale down a recipe.

The keys to scaling a recipe

You can scale up or down in either of two ways: by percentage or by ratio-proportion. Here's a quick rundown of how to use each method:

✔ **Scaling by percentage:** Find the percentage increase or decrease between the number of portions given and the number of portions you want, and apply the percentage to each ingredient. This method is fast, especially when the percentage increase or decrease is a nice round number.

✔ **Scaling by ratio-proportion:** Use the ratio of the number of portions given to the number of portions you want and apply it to the ratio of the given amount of each ingredient to the amount you need. This method shows you the clearest picture of the relationship between old and new amounts.

Scaling in action

To see how scaling works, consider a simple spaghetti meal that serves 4. It has only 3 major ingredients:

✔ 16 ounces uncooked spaghetti

✔ 24 ounces spaghetti sauce, made from crushed tomatoes

✔ 6 ounces Parmesan cheese

You'll want to add garlic, thyme, oregano, basil, marjoram, bay leaf, salt, and pepper as well, but the amounts of these ingredients are based on your personal preference.

Now say that you're having 12 people over on Saturday night. Since the recipe makes 4 servings, you need to multiply the ingredients by 3 to get the 12 serving you need (4 servings × 3 = 12 servings).

Or maybe only you and another person are eating, so you want 2 servings instead of 4, especially if you hate leftovers. How much of the ingredients do you need now? In this case, the solution — halving the recipe or dividing all the ingredients by 2 — may be apparent. With a simple recipe like this one, you see that mental math produces fast answers. (See Chapter 4 for more calculations you can do in your head.)

But suppose that the solutions don't just pop into your head or that the recipe is much more complex, with many more ingredients needed in much more precise quantities. That's when the two scaling methods can come in handy. I use a simple recipe in the following examples, but the math works with complicated recipes, too. After a while, you'll be able to do the math in your head and just write down the revised quantities on the recipe.

Scaling up and down by percentage

To scale up by percentage, do the following.

1. **Calculate the percentage you need to increase each ingredient.**

   Use this calculation:

   $$\text{percentage} = \frac{\text{desired servings}}{\text{recipe servings}} \times 100$$

   $$\text{percentage} = \frac{12}{4} \times 100$$

   $$\text{percentage} = 3.00 \times 100$$

   $$\text{percentage} = 300 \text{ percent}$$

2. **Multiply each ingredient by the percentage.**

   In this example, the percentage is 300 percent, so you multiply each ingredient by 3.00:

   Spaghetti = 16 oz × 3.00 = 48 oz

   Sauce = 24 oz × 3.00 = 72 oz

   Parmesan = 6 oz × 3.00 = 18 oz

Now you know that you need 48 ounces of spaghetti, 72 ounces of spaghetti sauce, and 18 ounces of Parmesan cheese.

To scale down by percentage, do the following.

1. **Calculate the percentage you need to decrease each ingredient.**

   Use this calculation:

   $$\text{percentage} = \frac{\text{desired servings}}{\text{recipe servings}} \times 100$$

   $$\text{percentage} = \frac{2}{4} \times 100$$

   $$\text{percentage} = 0.50 \times 100$$

   $$\text{percentage} = 50 \text{ percent}$$

2. **Multiply each ingredient by the percentage.**

   In this example, you multiply by 50 percent, or 0.50:

   Spaghetti = 16 oz × 0.50 = 8 oz

   Sauce = 24 oz × 0.50 = 12 oz

   Parmesan = 6 oz × 0.50 = 3 oz

So to feed 2 rather than 4 people, you need 8 ounces of spaghetti, 12 ounces of spaghetti sauce, and 3 ounces of cheese.

Scaling up or down by ratio-proportion

To scale up by ratio-proportion, follow these steps:

1. **Set up a ratio-proportion.**

   In this example, you are scaling up from 4 servings to 12 servings.

   $$\frac{\text{known servings}}{\text{desired servings}} = \frac{\text{known ingredient amount}}{\text{desired ingredient amount}}$$

   $$\frac{4}{12} = \frac{16 \text{ oz (spaghetti)}}{x \text{ oz (spaghetti)}}$$

2. **Cross-multiply and solve.**

   $4x = 192$

   $x = 48$ oz

   The answer is 48 oz.

3. **Perform the same calculation for the other ingredients in the recipe.**

To scale down by ratio-proportion, follow these steps:

1. **Set up a ratio-proportion.**

   In this example, you are scaling down from 4 servings to 2 servings.

   $$\frac{\text{known servings}}{\text{desired servings}} = \frac{\text{known ingredient amount}}{\text{desired ingredient amount}}$$

   $$\frac{4}{2} = \frac{16 \text{ oz (spaghetti)}}{x \text{ oz (spaghetti)}}$$

2. **Cross-multiply and solve.**

   $4x = 32$

   $x = 8$ oz

   The answer is 8 oz.

3. **Do the same calculation for the other ingredients in the recipe.**

If you like to cook (as I do), you can add to the fun by knowing the history of the dishes you make. For example, the first Italian cookbook with a recipe for tomato sauce was *Lo Scalo alla Moderna*, published in 1692.

Using Math to Buy and Cook Smart

Use some smart kitchen math to figure out per-serving costs exactly. You may find that a meal you thought to be costly isn't really expensive at all.

You can also use math to get more value from food (that is, save money) when you look at the advantages of buying food in bulk.

Calculating per-serving costs

The process for calculating per-serving costs of a meal is simple:

1. **Add up the costs of all the ingredients that make up a dish to give you the total cost of the dish.**

2. **Divide the total cost of the dish by the number of servings.**

3. **Repeat Steps 1 and 2 to figure the costs of each of the other dishes in the meal.**

4. **Add up the per-serving cost of each dish to get the per-serving cost for the meal.**

Here's an example: Suppose that you want to calculate the per-meal cost of a low-cost, three-item spaghetti dinner that serves 4 (refer to the earlier section "Scaling in action"). For this example, I've added a loaf of pre-buttered garlic bread. Also, the spaghetti is the cheapest in the store, the sauce is pre-made, and the Parmesan cheese is pre-grated in the lowest-priced container sold. The price of each of the ingredients is as follows:

- ✔ 16 ounces of uncooked spaghetti, $1.00

- ✔ 24 ounces of spaghetti sauce, $2.50

- ✔ 6 ounces of Parmesan cheese, $2.00

- ✔ 1 loaf of garlic bread, $3.00

With the prices at hand, follow these steps:

1. Add the cost of the spaghetti ingredients.

total cost = ingredient 1 + ingredient 2 + ingredient 3

total cost = $1.00 + $2.50 + $2.00

total cost = $5.50

2. Divide the total cost ($5.50) by the number of servings.

$$\text{per-serving cost} = \frac{\text{ingredient cost}}{\text{number of servings}}$$

$$\text{per-serving cost} = \frac{\$5.50}{4}$$

per-serving cost = $1.375

The answer is about $1.38 per serving.

3. Divide the cost of the garlic bread ($3.00) by the number of servings.

$$\text{per-serving cost} = \frac{\text{ingredient cost}}{\text{number of servings}}$$

$$\text{per-serving cost} = \frac{\$3.00}{4}$$

per-serving cost = $0.75

The answer is $0.75 per serving.

4. Add the cost of the items (spaghetti and garlic bread).

per-serving cost (whole meal) = item 1 + item 2

per-serving cost (whole meal) = $1.38 + $0.75

per-serving cost (whole meal) = $2.13

The answer is about $2.13 per serving.

Restaurants always calculate per-serving costs, and you can find online calculators that can help you. A good one is http://www.free-online-calculator-use.com/restaurant-food-cost-calculator.html. Also, want some ideas for inexpensive meals? Run an Internet search for "Meals for under $20." You'll get lots of hits. For that matter, plenty of sites list meals under $10.00. Many of these meals are one-pot meals, but you can find other types, too.

Taste by the ton: Buying in bulk

If you like or use some foods a lot, you may be able to save money by buying in bulk. Although buying in bulk can often save you a bundle, bulk prices aren't always a sure way to savings. The trick is in knowing when the bulk item is a better deal.

Fortunately, the math for comparing bulk versus non-bulk prices is easy. You simply calculate the per-ounce or per-pound cost of each product and then compare the answers. With this method, you can figure cost savings right at home, once you know a few prices. (To figure cost savings while you're shopping, jump to Chapter 5.)

You used to be able to find bulk food only in health food stores, but now you can find bulk bins in all major supermarkets, including giant national retailers. You can also find bulk bargains in specialty markets, such as grocery outlets and restaurant supply stores. The bulk bins are great for buying grains, pastas, rice, nuts, and cereals. If you're fortunate enough to live in or near a Latino or Hispanic community, you'll find big savings in big bags of beans and rice at the local *mercado.* The same reasoning applies to Indian or Pakistani items in England, or North African items in France.

Say you want to compare the price of a little can of kidney beans to a big one. The little can weighs 15.25 ounces and sells for $1.19. The big can weighs 108 ounces and sells for $4.69. Follow these steps to determine which is the better deal:

1. **Determine the per-ounce price of the little can.**

$$\text{per-ounce price} = \frac{\text{price}}{\text{weight (oz)}}$$

$$\text{per-ounce price} = \frac{\$1.19}{15.25}$$

$$\text{per-ounce price} = \$0.078$$

   Each ounce of kidney beans from the little can costs about $0.08.

2. **Determine the per-ounce price of the big can.**

$$\text{per-ounce price} = \frac{\text{price}}{\text{weight (oz)}}$$

$$\text{per-ounce price} = \frac{\$4.69}{108}$$

$$\text{per-ounce price} = \$0.0434$$

   Each ounce of kidney beans from the big can costs a little over $0.04.

When is less more?

Math isn't subjective, but food is. Quality and the perception of quality can improve your satisfaction and *also* save you money. For example, if you want to serve your friends a nice steak dinner, you may first think that the best option is to serve big, fat 12-ounce top sirloin steaks. Well, think again.

Even bargain sirloin can cost $6.00 per pound. A 12-ounce steak is three-quarters (3/4) of a pound (a pound has 16 ounces; refer to Table 6-1). Using a calculator (or doing some mental math), you can determine that each steak costs $4.50. Filet mignon is $16.00 per pound. A 4-ounce filet is one-fourth (1/4) of a pound. Therefore, each 4 ounce steak costs $4.00. So you *save money* by serving filet mignon instead of top sirloin, because you pay $4.00 per steak instead of $4.50!

It's obvious that the savings is derived from reducing the tonnage of the steak, which is a good thing, because 3 ounces is currently the recommended, healthy portion size for steak. Therefore, serving a 4-ounce portion is generous. Plus filet mignon tastes better, has the reputation of being more "high class" than top sirloin, and will impress your friends.

3. Compare the two prices.

Compare 4 cents per ounce for the big can with 8 cents per ounce for the little can. The big can is a much better deal, because the per-ounce price is only half as much.

You can calculate dry products, such as white rice and beans, the same way. For example, if a 1-pound bag of white rice sells for $1.69 and a 20-pound bag of white rice sells for $11.98, the math is the same. The 1-pound bag costs about $0.10 per ounce, and the 20-pound bag costs about $0.04 per ounce. If you want to compare pounds (which is a little faster), the 1-pound bag costs $1.69 per pound, and the 20-pound bag costs about $0.60 per pound.

Chapter 7

It Does a Body Good: Math for Health and Well-Being

*Y*our health is important, and few people would dispute that statement. If you're healthy, you can take advantage of all of life's opportunities. If you're not healthy, you have some rough sailing ahead.

What is "health," anyway? The World Health Organization (WHO) has defined health as "a state of complete physical, mental, and social well-being and not merely the absence of disease or infirmity." Ho! That's a big order to fill!

Real-life math can help you get healthy and stay healthy. If you apply some basic arithmetic to the components of health — nutrition, weight management, exercise, and taking medications — you'll know whether you're on track or need to tweak your strategy a bit. In this chapter, you look at key elements of health and the accompanying math.

Figuring Your Nutritional Needs

To be at optimum health, you need to make sure you're getting the right amount of nutrients in what you eat. Each nutrient (protein, carbohydrates, and fat, as well as vitamins and minerals) has a positive impact. Of course, nutrients satisfy your appetite and give you caloric energy, but they do a lot more than that. Here's a quick rundown of the good stuff nutrients do for your body:

- **Protein:** Proteins are one of the building blocks of body tissue. Proteins maintain cell shape and are part of organized tissues (otherwise, you'd look like an amoeba). They are also needed for cell adhesion (think of white blood cells sticking to blood vessel walls to fight infection) and just plain growth. Proteins can come from animal sources (beef, chicken, fish, eggs, milk, and cheese) or vegetable sources (beans, quinoa, soy, lentils, and tempeh).

  Proteins should make up about 10 percent of your daily food intake. That's about 200 calories per day for a 2,000 calorie per day diet.

- **Carbohydrates:** Carbs are a great source of energy. They're also prominent in raising blood glucose levels. The *complex carbohydrate* world is where you find pasta, bread, and cereals. The *simple carbohydrate* world is the home of sugar.

  Complex carbs should make up about 60 percent of your daily calorie intake. Complex carbs should make up most of your daily carbohydrate intake. Simple carbs should be an occasional treat.

- **Fats:** Fat provides energy, helps your body absorb certain nutrients, and keeps your cells healthy. Fats can come from animals (butter and lard) or vegetables (peanut oil, coconut oil, and the ever-popular olive oil). There are saturated fats, unsaturated fats, and monounsaturated fats (of which olive oil is best known). Trans fats have a bad reputation because they increase your risk of getting coronary heart disease.

  A healthy diet includes a little less that 30 percent fat.

- **Vitamins and minerals:** Called micronutrients because your body needs small amounts, this class of nutrient includes vitamin A, the B vitamins, sodium, magnesium, and a whole host of others. Each has a particular function, but all of them together help ensure your health and well-being. To find out how much of each of these nutrients you need daily, head to the later section "Following the recommended daily allowance."

Proper nutrition requires that you take in the basic nutrients, but you *also* need to consume dietary fiber and micronutrients. And that's basically all there is to it. In the following sections, I tell you how to use real-life math to make sure you're getting the nutrients you need.

Read the label, Mabel: Nutrition facts

Many countries (including the United States, Canada, Australia, New Zealand, and those in the European Union) require labeling on packaged food. This label (called a *Nutrition Facts label* in the United States) is filled with useful information.

Learn to read the food label. You can use the information to see how many calories and which and how much of key nutrients (protein, fats, and carbohydrates) and micronutrients (vitamins and minerals) the product has. Figure 7-1 shows two nutrition labels for sausages.

Nutrition Facts	
Serving Size 2 oz (56g)	
Servings Per Container about 7	
Amount Per Serving	
Calories 170	Calories from Fat 140
	% Daily Value*
Total Fat 15g	23%
Saturated Fat 5g	26%
Trans Fat 0g	
Cholesterol 35mg	12%
Sodium 570mg	24%
Total Carbohydrate 2g	1%
Dietary Fiber 0g	0%
Sugars less than 1g	
Protein 7g	
Vitamin A 0% • Vitamin C 0%	
Calcium 0% • Iron 2%	
*Percent Daily Values are based on a 2,000 calorie diet.	

INGREDIENTS: PORK, WATER AND LESS THAN 2% OF THE FOLLOWING: SALT, CORN SYRUP, DEXTROSE, SPICES, BEEF, SODIUM PHOSPHATE, MONOSODIUM

Nutrition Facts			
Serving Size 1 Link (71g)			
Servings Per Container 4			
Amount Per Serving			
Calories 200	Calories from Fat 160		
	% Daily Value*		
Total Fat 18g	28%		
Saturated Fat 6g	30%		
Cholesterol 45mg	15%		
Sodium 630mg	26%		
Total Carbohydrate 1g	1%		
Dietary Fiber 0g	0%		
Sugars 1g			
Protein 10g			
Vitamin A 0% • Vitamin C 0%			
Calcium 0% • Iron 4%			

*Percent Daily Values (DV) are based on a 2,000 calorie diet. Your Daily Values may be higher or lower depending on your calorie needs.

	Calories:	2,000	2,500
Total Fat	Less than	65g	80g
Sat Fat	Less than	20g	25g
Cholesterol	Less than	300mg	300mg
Sodium	Less than	2,400mg	2,400mg
Total Carbohydrate		300g	375g
Dietary Fiber		25g	30g

INGREDIENTS: Pork, Water, Beer, With 2% or less of the following: Sea Salt, Sugar, Natural Spices (including black pepper, white pepper, mace, cardamon, mustard).

Illustration by Wiley, Composition Services Graphics

Figure 7-1: Reading nutrition labels.

The label does most of the real-life math for you. Among other things, the "% Daily Value" column shows what portion of recommended daily values the item contains for a 2,000- to 2,500-calorie-a-day diet. (Look at the footnote on the label to see the daily calorie intake.)

Pay particular attention to the following information:

- ✔ **Serving size and number of servings in a package:** This info tells you how large (or small) a serving is. Pay close attention because all the other information on the label is based on a *single serving*.

- ✔ **Calories per serving:** When you look at the calories per serving amount, pay attention to the number of servings in the package or the serving size. Many packages that appear to be single servings (like a 16-ounce bottle of soda, for example) actually have multiple servings in them.

You will be shocked (shocked!) to learn that some product manufacturers play fast and loose with portion sizes to make their products seem less calorie-laden. Take a look at Figure 7-1 again. The right-hand label is for a package containing 4 bratwursts. The links are 5 inches long and weigh 2.5 ounces each. The serving size is stated as "1 link" — pretty straightforward. The left-hand label provides info for a package that contains four 8-inch sausages. Each sausage weighs 7 ounces. But the serving size is 2 ounces! In other words, a single serving is 2/7 of a whole sausage, which works out to be a 2.29-inch piece of sausage. This is where most people trip up (and who can blame them?) in accurately figuring how many calories or nutrients they're really getting.

When what appears to be a single-serving package contains multiple servings, you can determine the total calories by multiplying the "Amount Per Serving" calories by the "Servings Per Container" number. And here's one more tip. That *standard serving size* for a soft drink is 8 fluid ounces, no matter what the label says.

- ✔ **The percentage of the macro- and micronutrients a serving gives you.** With this info, you can determine where the food falls on the "how healthy is it for you" scale. The percentages tell you if you're headed for trouble in the world of protein and fat, but especially fat. As a general rule, food high in saturated fats, cholesterol, and sodium are less healthy.

If you ever want to get a real handle on calories and serving sizes, buy a digital kitchen scale. The scale is great fun, and you can easily weigh out any quantity of any food you want to eat.

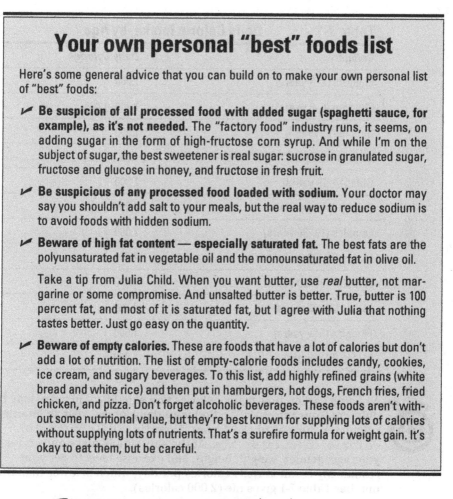

Your own personal "best" foods list

Here's some general advice that you can build on to make your own personal list of "best" foods:

✔ **Be suspicion of all processed food with added sugar (spaghetti sauce, for example), as it's not needed.** The "factory food" industry runs, it seems, on adding sugar in the form of high-fructose corn syrup. And while I'm on the subject of sugar, the best sweetener is real sugar: sucrose in granulated sugar, fructose and glucose in honey, and fructose in fresh fruit.

✔ **Be suspicious of any processed food loaded with sodium.** Your doctor may say you shouldn't add salt to your meals, but the real way to reduce sodium is to avoid foods with hidden sodium.

✔ **Beware of high fat content — especially saturated fat.** The best fats are the polyunsaturated fat in vegetable oil and the monounsaturated fat in olive oil.

Take a tip from Julia Child. When you want butter, use *real* butter, not margarine or some compromise. And unsalted butter is better. True, butter is 100 percent fat, and most of it is saturated fat, but I agree with Julia that nothing tastes better. Just go easy on the quantity.

✔ **Beware of empty calories.** These are foods that have a lot of calories but don't add a lot of nutrition. The list of empty-calorie foods includes candy, cookies, ice cream, and sugary beverages. To this list, add highly refined grains (white bread and white rice) and then put in hamburgers, hot dogs, French fries, fried chicken, and pizza. Don't forget alcoholic beverages. These foods aren't without some nutritional value, but they're best known for supplying lots of calories without supplying lots of nutrients. That's a surefire formula for weight gain. It's okay to eat them, but be careful.

Figuring out your ideal daily calorie intake

Your ideal daily caloric intake varies with your age, activity level, and other factors. Table 7-1 lists ages and the *average* calorie intake for each group (keep in mind that what you need may vary). These numbers come from the U. S. Department of Agriculture and assume an activity level of *30 minutes or less* per day. (That low activity, sadly, seems to be the trend in America.)

Table 7-1	Daily Calorie Intake, by Age
Category	*Daily Calories*
Children (2–3 years)	1,000
Children (4–8 years)	1,200–1,400
Girls (9–13 years)	1,600
Boys (9–13 years)	1,800
Girls (14–18 years)	1,800
Boys (14–18 years)	2,200
Females (19–30 years)	2,000
Males (19–30 years)	2,400
Females (31–50 years)	1,800
Males (31–50 years)	2,200
Females (51+ years)	1,600
Males (51+ years)	2,000

As I note earlier, how many calories you should eat in a day may be different from any chart. To find your daily intake, visit an Internet calculator, such as http://www.freedieting.com/tools/calorie_calculator.htm. At the site, simply make entries for your age, gender, weight, height, and exercise level to get your results. My result is 1,834 calories per day, which is lower than the number Table 7-1 gives me (2,000 calories).

Following the recommended daily allowance

The Dietary Reference Intake (DRI) is a set of recommendations from the U.S. National Academy of Sciences that lists about 29 micronutrients, ranging from vitamin A to zinc, and tells you how much of each you need daily. Both the U.S. and Canada use the DRI. If you're meeting the DRI recommendations, you're probably eating healthy.

You may know the DRI by another acronym: RDA, which stands for *Recommended Daily Allowance*. Same idea and essentially the same list with a different name.

Your everyday-math task is to be familiar with the list and apply it when you're reading food labels. It's an analysis and judgment job. (After all, you can't be doing arithmetic all the time.) You can find the list online at in many places. The table at the USDA website is too hard to read; instead, visit `http://en.wikipedia.org/wiki/Dietary_Reference_Intake`.

Calculating Calories

The *food calorie* (also known as the *calorie* or the *kcal*) is the measure of the energy in the food you eat, and it comes from the protein, carbohydrates, fat, and alcohol in a food item. That energy, when released, powers your body and your brain. It's the cornerstone of physical health.

Counting calories in your food

You know that different foods contain different amounts of calories. For example, an ordinary slice of bacon has a lot of calories (about 43) and a medium cucumber has few calories (about 24).

Calorie math is based on what you eat and how much you eat. It's a simple calculation: *total calories = number of servings × calories per serving*. A thick strip of bacon, for example, has 61 calories, mostly from fat. If you have 4 strips of bacon at breakfast, that's 244 calories. If you love bacon and eat a pound at a time, you've consumed 2,440 calories.

To find the calories in a complete meal, you determine the number of calories for each item and then add them all together.

Here are several pointers to keep in mind as you calculate the number of calories you're eating:

✔ For many foods, you can easily determine calories by reading food labels. But that doesn't work in the meat department at your supermarket. When it comes to beef, chicken, and fish, you have to do some calculating or look up the food on the Internet.

Honey, does this country make my butt look big?

The United States leads the world in obesity. Some health experts call it an epidemic, and the problem applies to children as well as adults. There are other "overweight countries," too.

The following table compares obesity in selected countries. Use your analysis skills to draw conclusions.

Country	Obesity Rank	Percent Obese
U.S.	1	30.60
Mexico	2	24.20
UK/Wales	3	23.00
Germany	14	12.90
France	23	9.40
Italy	25	8.60

By the way, in the U.S. Mississippi leads with 34.9 percent obesity in adults, and Colorado is at the tail end, with 20.7 percent obesity in adults.

Who cares? You should, because obesity contributes to all kinds of illness, including coronary heart disease and diabetes. First you get fat; then you get dead. (And this comes from an author who is overweight and fighting it every day.)

Math alert! Statistics are measurements made on large populations. The numbers for countries or states don't account for individual height, health, eating disorders, or genetics. See Chapter 3 for more on statistics.

I found the world's best calorie and nutrition information at http://nutritiondata.self.com. The site has thousands of items (including fast food items) and allows you to build your own personal list of foods you frequently eat.

✔ You can find calorie charts on the Internet that list the number of calories in a whole range of foods. Just do some multiplication to convert the amount of food shown in a chart to the amount you really eat.

✔ Pay attention to serving sizes. If a standard serving of ice cream is a half cup but you fill a cereal bowl, you're eating more than a single serving. Adjust your "Number of servings" accordingly to get an accurate calorie count.

Calorie algebra

When you eat, you're taking in nutrients in the form of protein, carbohydrates, and fat. Alcohol has calories but isn't accepted as a nutrient. Food also includes micronutrients. See the section "What's in the diet?" for details on those. The different nutrients (and alcohol) have a different number of calories:

✔ **Proteins:** About 4 calories per gram.

✔ **Carbohydrates:** About 4 calories per gram.

✔ **Fats:** About 9 calories per gram.

✔ **Alcohol:** About 7 calories per gram.

Don't let calorie counts baffle you. For example, a 4-ounce steak has more calories than 4 ounces worth of protein does. Why? Because a steak isn't all protein. A steak has plenty of fat, and that sends the cal count up. You can find out more about nutrients and how to make sure you get enough of them in the earlier section "Figuring Your Nutritional Needs."

Managing your weight with math

Your body knows how many calories it needs in a day, and it will tell you. How? By gaining or losing weight. If you're getting too many calories, your body stores the excess calories as body fat ("Once on the lips; forever on the hips"), and you gain weight. If you're getting too few calories, your body uses body fat to get the energy it needs, and you lose weight.

To figure out how many calories per day you need, use an Internet calorie calculator. Visit http://www.freedieting.com/tools/calorie_calculator.htm

Once you know how many calories your body needs in a day to maintain your current weight, you can decrease that number to lose weight or increase it to gain weight.

Of course, if your goal is to lose weight, you want do so safely, and that means losing no more than 1 to 2 pounds in a week. Here's what you need to know to manage your rate of weight loss:

✔ One pound (454 grams) of body fat stores about 3,500 calories.

✔ To lose 1 pound a week, you need to reduce the total number of calories you eat in a week by 3,500. To lose 2 pounds a week, reduce your weekly calorie intake by 7,000 calories.

✔ To make your weekly weight loss goal, divide the number of fewer calories you need to eat in a week by 7, the number of days in the week. This tells you how many fewer calories you need to eat each day.

$$\text{reduced calories per day} = \frac{\text{reduced calories per week}}{7}$$

$$\text{reduced calories per day} = \frac{3,500}{7}$$

$$\text{reduced calories per day} = 500$$

To lose 1 pound a week, you need to eat 500 fewer calories in a day. To lose 2 pounds, you need to eat 1,000 fewer calories a day.

If you find yourself facing a class reunion or wedding a few pounds heavier than you'd like to be, you can figure how many pounds you need to lose in the time left by dividing the weight you want to lose by the number of weeks you have to lose it.

$$\text{weight loss per week} = \frac{\text{pounds to lose}}{\text{number of weeks}}$$

$$\text{weight loss per week} = \frac{10}{5}$$

$$\text{weight loss per week} = 2$$

Be careful about crash diets. Your doctor won't like the idea, and many people who lose weight fast gain it back fast. Losing several pounds slowly is more sensible than losing them quickly. If you know you have an event coming up that you want to look your best for — and if "looking your best" means losing weight — start early enough that you can reach your goal without going on a crash diet. As I mention earlier, 1 to 2 pounds a week is a safe rate of weight loss.

Comparing your current weight to your goal weight

Weighing too little isn't good, but the vast majority of people weigh too much. If you're in the majority, you're probably trying to shed a few pounds. If you want to get down to your ideal weight, you need to know two things:

✔ **Your ideal weight:** Your *ideal weight,* or *goal weight,* is a generalized number that depends on your gender, height, and frame size. You can consult a chart or an Internet calculator. (You'll find a good calculator at http://www.halls.md/ideal-weight/body.htm.)

When you find your goal weight, don't panic! All goal weight charts and calculators will make you feel as though they're written for half-starved supermodels. Talk with your doctor. The doctor likes lean patients (because they have fewer health problems), but he or she won't suggest an insane goal weight.

✔ **Your current weight:** Oh, no! Not the scale! Yes, the scale. A good scale tells you how much you weigh. No math is involved; you merely have to read numbers correctly. So step on the darned thing and read your weight. (By the way, if possible, get a nice digital scale; many of them are accurate within 0.2 pound.)

Staying on point

Weight Watchers is a famous program that calculates food in "points." You're allowed so many points per day, plus some additional weekly points. The point method is intended to be easier than counting calories, and the program is intended to promote good eating habits and more exercise.

Point values are essentially designed to produce a 1,000 calorie daily deficit in your eating (7,000 calories per week), which is likely to result in a weight loss of 2 pounds per week (see the preceding section).

Here are the key point calculations:

✔ The points a food offers are based on a fairly complex proprietary formula, so you rely on Internet listings and the numbers on the branded food items the program sells in stores. (The point values have quite a range. For example, a banana split is 19 points, while bamboo shoots are 0 points.)

✔ Calculating the points you're allowed requires the use of official calculators, but you can find other point calculators on the Internet. You enter your age, weight, height, exercise level, and number of pounds you want to lose. The calculator assigns you 26–71 points to consume each day. You also get 49 additional "weekly points" to use each week.

✔ Your daily calculations are simple. When you know your daily and weekly points, just write down the point values for what you eat and add 'em up. Compare the points for what you eat to the points you are allowed to see whether you're staying on track or veering off the path, so to speak.

Remember: Point counting and calorie counting have limits. They don't indicate nutritional value. Develop healthy eating habits.

With these two numbers (current weight and goal weight), you simply subtract to see how much you need to lose (or gain). When your current weight is greater than your goal weight, use this formula (in this example, the current weight is 189 pounds and the goal weight is 160 pounds):

amount of weight to lose = current weight − goal weight

amount of weight to lose = 189 − 160

amount of weight to lose = 29

When your current weight is less than your goal weight, you use this formula (here, current weigh tis 138 and goal weight is 150):

amount of weight to gain = goal weight − current weight

amount of weight to gain = 150 − 138

amount of weight to gain = 12

In the first case, you might consider losing about 29 pounds. In the second case, think about gaining 12 pounds.

Diet versus dieting

Your *diet* is a broad term for everything you eat. It's largely influenced by culture. Your diet is a composite of your individual choices, too. You might choose a *vegan* diet and eat no dairy or meat, nor anything produced by an animal (like honey); a *pescetarian* diet , which includes fish but no meat; an *ovo-lacto vegetarian* diet, in which you eat eggs and diary, but no meat; or any number of other specialized diets.

But for most of history, people have eaten what food they could get. The result has been a number of reasonably healthy nations. For example, many people in Azerbaijan (a relatively poor country) live to 100 or older, possibly because they eat a lot of yogurt, a lot of vegetables, and relatively little meat.

Dieting is different. Dieting is a conscious plan to gain weight (sometimes) or lose weight (usually).

Calculating BMI

The *body mass index* (BMI), a measurement that's been around for over 150 years, is an approximation of how much body fat you have. Your BMI is a really handy number, and it's easy to calculate. The basic idea is that, if you have a high BMI, you're carrying around too much fat. A BMI over 30 means you fit the definition of obese.

To calculate your BMI, you divide your weight by the square of your height.

To calculate BMI with metric units:

$$BMI = \frac{\text{weight in kg}}{(\text{height in meters})^2}$$

$$BMI = \frac{92.08}{(1.752)^2}$$

$$BMI = \frac{92.08}{3.069}$$

$$BMI = 30.00$$

In American units, the formula is a little different, because it must factor in metric-to-American conversions.

$$BMI = \frac{\text{weight in lbs}}{(\text{height in inches})^2}(703)$$

$$BMI = \frac{203}{(69)^2}(703)$$

$$BMI = \frac{203}{4761}(703)$$

$$BMI = 29.97$$

Don't want to do the math yourself? You can go online and get the number from a BMI calculator or BMI charts (easy math, right?). Visit http://www.webmd.com/diet/calc-bmi-plus. The calculator not only gives you your BMI, but it also gives you a number for your body shape (waist-to-height ratio).

Exercise Math

Exercise? What? I'd rather not. But some people — maybe even you — love it. Some, like my doctor (a mountain biker), even insist that exercise can become a wonderful addiction.

Doctors recommend exercise to help almost every health condition: obesity, diabetes, coronary heart disease, and osteoporosis, to name a few. And many people exercise to help lose weight. The object is to expend calories, that is, to "burn off" stored body fat in an activity.

These METs won't win a pennant: Metabolic equivalent of task

The *metabolic equivalent of task* (MET) is a relative measure of the energy cost of physical activity, which is a fancy way of saying it lets you compare the burn rate of different forms of exercise. Figure 7-2 shows the relative levels of effort of different kinds of exercise.

Activity	MET
Watching television	1
Desk work	1.8
Walking (1.7 mph)	2.3
Walking (2.5 mph)	2.9
Walking (3.0 mph)	3.3
Walking (3.4 mph)	3.6
Bicycling	4
Stationary bicycling	5.5
Jogging	7

Illustration by Wiley, Composition Services Graphics

Figure 7-2: MET chart for different activities.

The higher the MET value, the more energy (and therefore more calories) an exercise burns. From Figure 7-2, for example, you see that jogging expends 7 times the energy that watching television does. The "exercise value" of a MET chart is that you can find forms of exercise with higher burn rates (for example, stationary bicycling at the gym). The "monetary value" of a MET is that you can see what exercise is low-cost or no-cost (for example, walking and jogging).

The actual burn rate of any particular activity varies from person to person and depends on your current weight and fitness level, but the MET is *relative* to all the forms of exercise you do. No matter what your age or weight, you'll still burn more calories walking than watching TV.

Aerobic versus anaerobic exercise

You don't need to be a doctor of sports medicine to know about the two broad categories of exercise. *Aerobic* exercise consists of activities with repetitious movements over a relatively long period of time. Aerobic exercise is great for your respiration and heart function, and includes the treadmill and step aerobics.

Anerobic exercise consists of activities that require extreme expenditures of energy over a relatively short period of time. Anaerobic exercise builds strength and muscle mass, and includes weight training (also known as "pumping iron").

Down at the gymnasium, you can take in both kinds of exercise.

Figuring an activity's calorie burn rate

While MET values show the relative intensity of different forms of exercise, you will find greater value in knowing the absolute number of calories that are "burned." How many calories you expend per hour (called the burn rate) varies, based on your body weight. Figure 7-3 shows a sampling of calories used in 1 hour of cycling for people of different weights.

Activity (1 hour)	130 lb	155 lb	180 lb	205 lb
Cycling, <10 mph, leisure bicycling	236	281	327	372
Cycling, 14-15.9 mph, vigorous	590	704	817	931
Stationary cycling, light	325	387	449	512
Stationary cycling, very vigorous	738	880	1022	1163

Illustration by Wiley, Composition Services Graphics

Figure 7-3: Burn rates vary, based on intensity of the activity and body weight.

 You can find such charts on the Internet. To see a lengthy list of activities, visit http://www.nutristrategy.com/activity list4.htm; then just find the activity you like and find your weight. Write down the "calories per hour" number and figure how much time you will do the activity.

 As helpful as the such lists are, even more helpful are calculators that let you enter *your* data — weight, intensity level of activity, and amount of time you performed the activity — and then produce a report with the number of calories burned. Visit http://www.healthstatus.com/calculate/cbc and give it a try.

In the following list, I give approximate burn rates for a few activities:

- **Walking on a treadmill:** Walking on a treadmill not only improves fitness, but burns calories of body fat. The exact burn rate depends on your weight and metabolism. A person weighing 180 pounds treading at 4 miles per hour for 60 minutes burns about 421 calories.

- **Lifting weights:** Aside from the other benefits of weight training (building muscle or firming muscle), a 180-pound person lifting weights for 60 minutes burns 281 calories.

- **Sleeping:** You burn calories even while you're sleeping or resting. (Is that great news, or what?) The number of calories you expend depends on your weight and how long you sleep. For example, a 150-pound person sleeping for 8 hours burns 504 calories! That's 63 calories per hour.

Your friend, the pedometer

A *pedometer* is a small device that counts the number of steps you take when you walk. Everybody's step is a little different, so you do an initial calibration that enables the device to accurately "know" the length of *your* step. This formula is

step length = known distance ÷ number of steps

You set up for conversion between your step (measured in feet) and walking distance (measured in miles) by walking a known distance, such as 100 feet. You divide 100 by the number of steps on the pedometer to determine your step length (feet per step).

The first pedometers were mechanical. You attached the meter to your waist, and when you took a step, the counter registered it. You then later converted the count into miles "manually."

Say your step is about 3 feet. With the classic pedometer, you would later convert using the formula

miles = (number of steps ÷ step length) ÷ 5,280

Modern pedometers have silicon chips and store the initial calibration, so they do mileage calculations for you automatically.

Nowadays, you can get a *free* smartphone pedometer app that does everything. The app stores distance, time, speed, and calories burned. In addition (and this is really cool), the app uses the phone's GPS feature to show you a map of where you just walked. A typical app is "Pedometer FREE," available at Apple's iTunes Store, and there are similar apps at the Android store.

Being the Doctor at Home

When your spouse, your kid, an older relative, or a close friend gets sick, who they gonna call? Probably you. Acting out the role of doctor and dispensing medication at home requires great care and some easy math. The key techniques are to understand medication labels and to dispense the medications correctly.

Understanding medicine labels

Prescription and non-prescription medications have labels, which tell you a lot about the meds. These labels aren't really complicated, but at first they may seem so, because they contain so much legally required information, some of which looks to be in code. Figure 7-4 shows a typical medicine label.

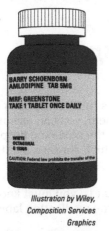

BARRY SCHOENBORN
AMLODIPINE   TAB 5MG

MRF: GREENSTONE
TAKE 1 TABLET ONCE DAILY

WHITE
OCTAGONAL
G 1500/5

CAUTION: Federal law prohibits the transfer of this

Illustration by Wiley,
Composition Services
Graphics

Figure 7-4: A typical medicine label.

The vast majority of medications come in the form of *tablets* (tabs) and *capsules* (caps). That term will be on the bottle.

The important information on the label is:

- ✔ **Name:** The brand name or generic name of the drug.

- ✔ **Amount of the active ingredient:** This info is usually shown in milligrams (mg): For example, aspirin comes in 325 mg tablets for adults and 81 mg tablets for children. Occasionally, the active ingredient is noted in micrograms (mcg).

- ✔ **Dosing:** This info tells you how much of the medication to take and with what frequency. For example, "1 tablet 4 times daily."

✔ **Other instructions:** If the medication should be taken at a particular time (morning, for example, or with food), that info will be on the label as well.

✔ **Other useful stuff:** The label also includes the manufacturer's name, expiration date, and pharmacy name.

In Figure 7-4, you see that the drug is amlodipine, the dosage is 5 mg, and the patient should take 1 tablet per day.

The entirety of real-life math for tablets and capsules is to *read* the label, *count* the tablets in a single dose, and *count* the number of times the med has been given in a day. The math seems simple, yet it's vital.

Splitting a tablet is okay, if the tablet has a score line down the middle. The score line lets you divide the tab evenly. Use a little device called a *pill splitter,* which you can get at any drugstore. *It's not okay to split capsules.*

For good measure: Dispensing liquid medications

Liquid medicines are dosed in milliliters (mL), teaspoons (tsp), and tablespoons (tbsp). To dispense them, you can use one of three low-cost tools: the medicine cup, the medicine dropper, and the measuring spoon (see Figure 7-5, which shows a medicine cup and a medicine dropper).

The math is "measurement math" — that is, you must know how much medicine to dispense, based on the medication label, and then measure it accurately.

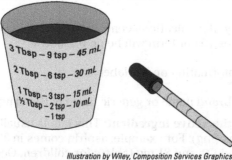

Illustration by Wiley, Composition Services Graphics

Figure 7-5: A medicine cup and medicine dropper.

✔ **Medicine cup:** Essentially, a medicine cup is a shot glass with various accurate measurements on it. It doesn't just allow for accurate measurement; it also does liquid volume conversions for you, showing the relationship between teaspoons, tablespoons, and milliliters.

✔ **Medicine dropper:** A medicine dropper is a glass or plastic tube that looks like an itty-bitty turkey baster, with a narrow end and a rubber "squeezer" at the other end. The barrel has milliliter (mL) marks. Often dispensing drops simply requires that you count the drops. This works for eye drops, ear drops, and nose drops. (As a bonus, it also works for food coloring when you're dyeing Easter eggs.) If you have to dose more than a few drops, you fill the dropper to the correct mark (say 0.5 mL or 1.0 mL), and then give it a gentle squeeze.

✔ **Measuring spoons:** Measuring spoons are precise measures, and you probably have a set in your kitchen. You typically use the teaspoon (5 mL) and tablespoon (15 mL) for dispensing meds. The best sets show both American units and metric units.

Although your Gorham sterling silver teaspoons are beautiful (at about $220.00 each), they aren't accurate for measuring medicines.

Converting tsp and tbsp to mL

Converting from teaspoons or tablespoons of liquid medications is easy if you know that 1 teaspoon equals 5 mL, 1 tablespoon equals 3 teaspoons, and 1 fluid ounce equals 2 tablespoons. Like much of real-life math, the calculation is simple, and you can easily do it in your head. But if you don't want to, here are the conversions:

1 teaspoon (tsp) = 5 mL

1 tablespoon (tbsp) = 3 teaspoons = 15 mL

1 fluid ounce (fl oz) = 2 tablespoons = 30 mL

Now you know what to do if you have to take 15 mL of cold medicine but have lost the measuring cup that so conveniently comes with the bottle. Just pull a tablespoon out of your kitchen drawer.

Converting drops to fluid ounces

Converting drops to fluid ounces and other units is a total waste of your math skills. And don't bother consulting the Internet, where you'll find wild variations in equivalents, for example 480, 354.88, 360.00, and 591.47 drops per fluid ounce. Smart math amounts to ignoring conversions you'll never use.

If you're in the hospital undergoing intravenous therapy ("on an IV"), drops are more precise. The tubing is calibrated to deliver 10, 15, or 20 drops/mL for adults. Children use different tubing that delivers 60 drops/mL. Don't worry. Your nurse knows how it works.

Chapter 8

Putting Geometry to Work at Home

In This Chapter

▶ Calculating areas and volumes for a variety of home projects

▶ Getting familiar with conversions so you purchase just what you need

You use math, especially geometry, around the house. All you need (for a start) is to know how to calculate the area of a rectangle (and maybe an occasional triangle or circle). I cover the general how-to's for finding areas in Chapter 2. In this chapter, I show you how easy the math can be when you tackle some common home maintenance and improvement tasks.

Note: Although I deal with specific scenarios in this chapter — like how to find the area of a flower bed so that you know how much mulch to buy — you can use the same techniques for other tasks. After all, the same principles — and math — apply whether you're seeding a lawn or carpeting a room.

Calculating Your Way to a Better Lawn and Garden

Taking care of the lawn, the flower beds, and the vegetable garden doesn't necessarily have to be tedious; it can be fun. Mowing the lawn was a chore when I was a kid, but the results were satisfying. And working with flower beds and a vegetable garden gives most people a lot of satisfaction.

Aside from mowing, digging, and planting, a lot of work around the home deals with spreading materials around — be they paint, concrete, or mulch. Each process takes a little math. And let's not forget that math can also answer the question that anyone who's ever pushed a lawn mower on a hot summer afternoon has pondered: Just how big *is* this yard?

Figuring how much seed you need

A great way to improve your lawn is by over-seeding. While fertilizing is important, over-seeding is a super way to make your lawn thick and weed-free. The question is, how much seed do you need to get the lush lawn you're striving for?

To find the answer, calculate the area of your lawn and calculate how many pounds of seed you need. For this example, assume that the lawn is a rectangle on a typical city lot.

Follow these steps:

1. **Calculate the area of the front lawn.**

   A typical city lot width is 65 feet wide. If your house is set back 25 feet from the street, the front yard's length is 25 feet. Now you have the numbers you need to figure area, using this formula: area = length × width.

   area (square feet) = length (feet) × width (feet)
   area (square feet) = 25 × 65
   area (square feet) = 1,625

   The area of the front lawn is 1,625 square feet. (Head to Chapter 2 for general info on how to calculate the areas of other shapes.)

2. **Calculate number of pounds of seed you need.**

   Seed companies make different claims about how much coverage a pound of seed gives you. It depends in part on the type of grass you want to plant. But for this example, suppose the seed company claims 600 square feet per pound.

   $$\text{seed needed (pounds)} = \frac{\text{area of your lawn (square feet)}}{\text{area covered by 1 pound (square feet)}}$$

   $$\text{seed needed (pounds)} = \frac{1,625}{600}$$

   $$\text{seed needed (pounds)} = 2.7$$

The answer is about 2.7 pounds, so you'll need a three-pound box or bag of seed.

 Some landscapers suggest using the same amount of seed for over-seeding as you would for a new lawn. Also, these calculations work the same way for fertilizing or seeding a new lawn, and the area calculation is useful for the costly option of bringing in sod.

Mulching math

Your beds will look nice with a fresh batch of mulch. Mulching math is different from seeding math because mulching involves depth. Here's how you determine how much mulch you need:

1. **Calculate the volume of the bed.**

   To mulch a 4 foot by 10 foot bed to a depth of 2 inches, first find the volume in cubic feet. To do so, multiply the length × width × height.

    When you do calculations, you need to make sure that all the values are represented in the same units. In this example, you know that the bed is 4 feet by 10 feet, so you need to put the depth in feet, too. You may recall that 2 inches is 2/12 (1/6) of a foot, which is also 0.167 feet.

   volume (cubic feet) = length (feet) × width (feet) × depth (feet)

   volume (cubic feet) = $10 \times 4 \times 0.167$

   volume (cubic feet) = 6.68

   The answer is 6.68 cubic feet. (You read more about calculating volume in Chapter 2.)

2. **Calculate how many bags you need.**

   To calculate, just divide the volume you need by the number of cubic feet in a bag. Where I live, redwood or cedar bark is typically sold in 3 cubic foot (CF) bags.

   mulch needed (bags) = $\dfrac{6.68}{3}$

   mulch needed (bags) = 2.23

   The answer is 2.23 bags. You can buy three bags and use the leftover bark in another bed.

 The calculations you use for seeding and mulching work the same way for applying compost or manure.

There are many mulches. In California, redwood bark is really popular. An eco-friendly mulch is rubber mulch, made from (I kid you not) ground-up tires. Unfortunately, math can't help you decide which mulch you prefer.

Planting the seeds of success — mathematically

A challenge to gardeners is figuring out how many plants you have space for. To get the answer, you need to know how large your plot is and how far apart the plants need to be. Say you love leaf lettuce so much that you want to fill a 2 foot by 4 foot raised vegetable plot with it. If you use a modern method of gardening, such as the French intensive method or Mel Bartholomew's square foot gardening method, the plants need only be spaced 4 inches apart.

1. **Measure the size of the area you want to plant.**

   As noted, in this example, assume you want to plant a 2 foot x 4 foot area.

2. **Figure how many plants you can fit in a foot; then multiply that number by the number of feet in a row and subtract 1.**

   Assume you want your plants to be 4 inches apart. Because 4 inches (the space between plants) is 1/3 foot, you get 3 plants per foot. And because you have 4 feet in each row, you calculate that you can fit 12 plants in each row, right? Not quite. Don't forget to subtract 1. If you don't subtract 1, the last plant in each row bumps up against the edge of the raised bed. Figure 8-1 illustrates the problem.

   plants in a row = length of row (feet) × number of plants per foot − 1

   plants in a row = (4 × 3) − 1

   plants in a row = 12 − 1

   plants in a row = 11

3. **Repeat Step 2 for the other dimension to determine the number of rows you have.**

   This time the calculation is (2 × 3) − 1, giving you 5 rows.

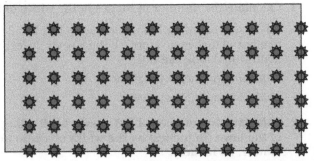

Illustration by Wiley, Composition Services Graphics

Figure 8-1: You need to adjust your calculations so that your rows of plant don't run into the edge of the bed.

4. **Multiply the number of plants in a row by the number of rows.**

   In the example, you have 5 rows and 11 plants in a row, so the correct answer is 55 plants.

   Note that real life math isn't always the same as classroom math. It often pays to make a sketch to help you visualize what you want to do.

Knowing how much you really mow

Mowing the lawn doesn't have any unusual math, right? Wrong! When you mow the lawn, you often mow about *twice* the area of the lawn. That's due to using the *half-pass technique*. With the half-pass technique, when you mow a strip of lawn, you don't move over a full width for the next row; instead, you move over about half a row. With this technique, you mow what amounts to a narrower strip, but the job's easier, and you don't miss any patches. Yes, it's really a full pass in length, but it's known as a half-pass because each pass covers only half the width of a row.

To find out what you're really cutting, you need to know the width of your lawn mower and the width of the yard. (This example uses a 22-inch rotary mower and a lawn that's 20 feet wide.) Then follow these steps:

1. **Calculate the number of inches in the width of the lawn.**

   To do so, you multiply the number of feet by 12 (the number of inches in a foot). If the lawn is 20 feet wide, the math would look like this:

   width of lawn (inches) = 12×20

   width of lawn (inches) = 240

2. **Calculate the number of full passes by dividing the width of the lawn (in inches) by the width of a pass (for this mower, 22 inches).**

   number of passes = $\dfrac{\text{width of lawn (inches)}}{\text{width of pass (inches)}}$

   number of passes = $\dfrac{240}{22}$

   number of passes = 10.9

   The answer is 11 full passes. That's what you'd cut if you didn't allow for half-passes.

3. **The number of so-called half-passes is about the same as the number of full passes. Add them together.**

   $11 + 11 = 22$

   You'll make 22 total passes over the lawn.

   Whether you share this information with the kid you pay to mow your lawn is entirely up to you.

Fixing Up the Place

You can make your home more attractive and useful by fixing it up, and if you own your own home, improvements often help its value. But most home maintenance or redecorating projects require some facility with math, whether you plan to tackle the jobs yourself or hire experts to do them for you. The next sections tell you how to do the calculations for a few common projects. As always, you can apply these calculations to other tasks.

Laying carpet

New carpeting does a room a world of good, but it usually makes a dent in your wallet, whether you're buying more economical brands or the best the market has to offer. The basis for carpeting costs is the area of the room you want to carpet. Mathematically, the task is pretty simple: Calculate an area (square feet) and then do a unit conversion from square feet to square yards, because carpeting is sold in square yards.

Follow these steps:

1. **Calculate the area of the room.**

   Say you're carpeting a large living area that's 18 feet by 24 feet. Multiply the length by the width:

   area (square feet) = length (feet) × width (feet)

   area (square feet) = 18 × 24

   area (square feet) = 432

   The area of the room is 432 square feet.

2. **Convert from square feet to square yards.**

   Carpet is usually sold by the square yard. One square yard is equal to 9 square feet. (Imagine a square 3 feet long and 3 feet wide.) To make the conversion, you divide the area in square feet by 9:

   area (square yards) = $\frac{432}{9}$

   area (square yards) = 48

   The area is 48 square yards.

To determine the cost of the carpet, you multiply the cost per square yard by the number of square yards. But don't forget that you'll be buying a carpet pad, as well, and, if you opt for professional installation, you'll have to pony up the per-yard cost for that, too.

Calculating paint amounts

Painting a room is easy. You need the right equipment, and of course, you should buy the right amount of paint.

The amount of paint you need is based on the square footage the paint must cover. And the best shortcut in calculating square footage is to take the height of the room (in feet) and multiply that by the room's perimeter.

Here's the painless way to calculate your paint requirements:

1. **Calculate the perimeter of the room.**

   Measure the room around the baseboards to get the perimeter. Alternatively, if your room is rectangular, take the length of the room and double it, take the width of the room and double it, and then add the two values together. Here's the calculation for a 9 foot by 12 foot rectangular room:

perimeter (feet) = 2 × length (feet) + 2 × width (feet)

perimeter (feet) = (2 × 9) + (2 × 12)

perimeter (feet) = 18 + 24

perimeter (feet) = 42

The perimeter is 42 feet.

2. **Calculate the area of the walls by multiplying the perimeter by the wall height.**

    In the United States, the standard height of a room is 8 feet. If your walls are 8 feet tall, multiply the perimeter by 8:

    area (square feet) = perimeter × wall height

    area (square feet) = 42 × 8

    area (square feet) = 336

    The wall area is 336 square feet.

Subtracting the areas of doors and windows from the total area usually isn't productive. Having a little paint left over is better than running out.

3. **Calculate the number of one-gallon cans of paint you'll need.**

    First-class paint claims that "one coat covers." For this next calculation, assume that's true. If you want to buy paint that covers is 400 square feet (sf) per gallon, you do the following calculation to see how many gallon cans you need:

    $$\text{number of cans} = \frac{336 \text{ (square feet to cover)}}{400 \text{ (square feet from 1 can)}}$$

    number of cans = 0.84

    The answer is about 0.84 cans. That's one 1-gallon can.

Pouring a patio

Pouring a concrete patio is a great exercise in practical math, and you get a nice patio for your efforts. The main task is to determine how much concrete you need. To find out, you calculate an area (in square feet), then volume (in cubic feet), and finally do a unit conversion (from cubic feet to cubic yards).

Most patios are rectangular, but if your patio has angles or curves, you can figure out the area by calculating each part and adding them up. Figure 8-2 shows typical shapes of parts of a patio.

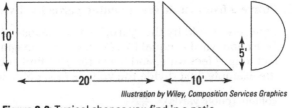

Illustration by Wiley, Composition Services Graphics

Figure 8-2: Typical shapes you find in a patio.

I cover the formulas for finding the areas of rectangles, triangles, and circles in Chapter 2. To find the area of a half circle, like the one shown in Figure 8-2, you just find the area of the full circle and divide it by 2.

To determine how much concrete you need to pour a cement patio, follow these steps:

1. **Calculate the area of the patio.**

   Assume you want to pour a 20 foot by 40 foot rectangular slab. (I have a friend who used this shape just outside his back sliding door.) You multiply the length by the width:

   area (square feet) = length (feet) × width (feet)

   area (square feet) = 20 × 40

   area (square feet) = 800

   The area is 800 square feet.

2. **Calculate the volume of the patio by multiplying the area by the depth (in feet).**

   A patio should be a minimum of 3.5 inches thick, but greater depth is better, especially if you're pouring the patio over expansive soil. For this example, imagine that you want your patio to be 6 inches thick, which just happens to be 0.5 feet (remember, you need to use the same units throughout your calculation).

   volume (cubic feet) = area (square feet) × depth (feet)

   volume (cubic feet) = 800 × 0.5

   volume (cubic feet) = 400

The volume is 400 cubic feet.

You can save a step by using the formula for volume:

volume = length × width × depth

3. **Convert from cubic feet to cubic yards by dividing by 27.**

Concrete is sold by the "yard," which means cubic yards. One cubic yard is equal to 27 cubic feet. (Imagine a cube 3 feet long, 3 feet wide, and 3 feet tall.) To find cubic yards in the example, just divide the volume (400 cubic feet) by 27.

$$\text{volume (cubic yards)} = \frac{400}{27}$$

$$\text{volume (cubic yards)} = 14.8$$

The volume is 14.8 yards.

Similar calculations work for decks, too. You determine the area to be decked and do a unit conversion — but in this case, you convert from area in square feet to board feet for buying the lumber.

Doing it yourself versus hiring a pro

What's the math (which amounts to cost savings) in "do it yourself" (DIY) work? The principle of making cost comparisons is universal throughout this book: Calculate the components of alternative choices, add the components up, and then compare the prices.

The big differentiator in DIY is the cost of labor. The calculation is simple:

total contractor cost = materials + contractor labor

total DIY cost = materials + your labor ($0.00)

cost savings = total contractor cost − total DIY cost

Although doing a task yourself may be satisfying and result in big savings, you've got to ask yourself some questions in a few key areas:

✔ **Special tools:** Do you have all the special equipment the job may require?

✔ **Skill:** Do you know what you're doing? Does it matter?

✔ **Quality of work:** Can you do the job well? Contractors are in business to deliver quality work.

✔ **Timing:** How long will it take you to do the job? A contractor can usually get the job done faster.

- ✔ **Interest and enthusiasm:** Do you *want* to do the job? If you like the work, rock on. If not, consider a contractor.

- ✔ **Value of your time:** Is it worth it? If you make $350.00 per hour as an attorney, common sense says that you'd be better off doing attorney work and leaving home improvements to a professional craftsperson.

- ✔ **Opportunity cost:** In economics, *opportunity cost* is defined as the value of the next best alternative when you choose to use your time (and money) for one activity instead of another. It's what you *don't* get to do if you choose an activity. For example, instead of doing a DIY project, would you rather go boating or write a novel?

Chapter 9

Math and Statistics around Town and on the Road

· ·

In This Chapter

▶ Calculating gas mileage, miles per gallon, distance, and more

▶ Figuring out how to calculate tips and divide a bill

▶ Choosing between flying and driving for vacation travel

▶ Using math to improve your odds of winning when you gamble

· ·

*N*o matter where you go, you'll want to know a few real-life math calculations. If you drive a car, you benefit from looking at regular driving and maintenance costs. If you like to go out to eat, math can help you calculate a tip or split a bill with ease. If you're planning a vacation, math can help you decide whether to go by plane, train, or automobile. And if your vacation should take you to Atlantic City, Las Vegas, or the fabled Monte Carlo, being able to calculate (or at least know) odds comes in mighty handy. The math-savvy gambler is a better gambler.

In this chapter, you find out what math to use in the places where you spend your time and money. The math works anywhere you go.

Automobile Arithmetic: Figuring Costs, Mileage, and More

Americans love to drive, and transportation drives American culture. People spend a lot of money for their vehicles. In fact, a car is usually the second biggest asset you own, after your home.

In addition to the cost of the car — and the interest you pay on the loan — cars have other expenses associated with them, too: fuel, insurance, and maintenance, for example. If you want to know how to figure how much your car loan is actually costing you, head to Chapter 10. But for other car-related calculations, keep reading.

It's a gas! Comparing fuel prices and mileage

The cost of gassing up isn't trivial. Fuel prices generally move in one direction — up. You feel it in your pocketbook, and the environment feels it, too. A country that burns a lot of hydrocarbons doesn't have nice air to breathe. China is a prime example of this. One obvious way to save money (and help the environment) is to be conscious of your gasoline mileage. Another way is to compare gas prices.

We call the instrument with "E" (for empty) and "F" (for full) a *gas gauge*, but *fuel gauge* would be a better term. Diesel fuel isn't gasoline, although an E85 blend is. Biofuels are something else. Get ready! As more electric cars take to the roads, "fuel" will be measured by a state of charge (SOC) meter.

Calculating mileage

You measure fuel economy in miles per gallon (mpg) in the United States. Other countries measure it in kilometers per liter (kpl). No matter how you measure it, driving a long distance using a minimum amount of fuel gives you a nice feeling.

Manufacturers tell you the average mileage of their vehicles, but they determine their figures under controlled driving conditions. To perform your own mileage calculations, you use a simple calculation: Divide the number of miles you drive on a full tank by the number of gallons in your tank.

For example, say your tank holds 20 gallons, and you drive 360 miles on that tank: 360 ÷ 20 = 18. You get 18 miles per gallon. The number is approximate, since most people won't drive until their tanks are completely empty before filling up. So consider driving until you've used up about 3/4 of the gasoline (15 gallons in the 20-gallon example). Repeat the calculations a few times, and you'll have a pretty good idea where your mileage stands. By the way, you can look up your tank's capacity in your car's owner's manual.

Car and truck advertising on the Internet sometimes buries fuel economy figures, because some luxury vehicles and big trucks are gas guzzlers. For the straight skinny on fuel economy, visit the U.S. Department of Energy website at http://www.fueleconomy.gov/.

Chasing down a good gasoline price

The cost of fueling is easy to calculate. Simply multiply the per-gallon price of the fuel by the number of gallons you buy (cost = gallons × price per gallon). In fact, you don't even have to do the calculation because the information is on the computerized receipt that the pump gives you at the end of fueling.

Here's a question that drivers frequently ponder, though: Should you drive across town for a lower gas price? The short answer is no. Never drive across town for a lower gas price. Gas prices usually vary by no more than maybe $0.05 (5 cents) per gallon. The math will show that you lose money by driving to save money.

Say you get 30 mpg and have to drive 5 miles out of your way to get a better gas price. That's a 10 mile round trip and consumes 1/3 of a gallon. If gas is selling for $3.60 per gallon (and that's low), you've used $1.20 worth of fuel. To make the trip worth your time, you need to save significantly more than $1.20.

If you have a 12 gallon tank (as I do), and the gasoline at the distant station is selling for $0.10 per gallon less than at your usual station, you end up breaking exactly even (12 gallons × 0.10 = $1.20). You don't save anything — and think about the time you wasted!

If you happen to shop for groceries at a major store that has a "gas island" at the far edge of the parking lot, buy your gas there. The gas price is usually lower, either because the store sets it low to attract shoppers or because you can get a per-gallon discount as a "Club Member."

The joys of a fuel log

If you are completely insane (as I am) you can compare gas prices over time by keeping a log. Use a spreadsheet program. The image here shows a detailed fuel log. The log not only shows you gas mileage, but it also shows you how local gas prices are trending and what you paid on a long drive. As a bonus, you can calculate days between fill-ups, changes in price, percent change in price, and average miles driven per day.

TANK	ODO	MILES	GALS	MPG	$/GAL	TOTAL $	DATE	Days Between Fill-Ups	Change in Price	Percent Change	Miles per Day	
33	104024	305	10.092	30.222	$4.019	$40.56	06/07/11	25	(0.21)	-4.97%	12.20	GV
34	104359	335	10.999	30.457	$3.959	$43.55	06/26/11	19	(0.06)	-1.49%	17.63	GV
35	104729	370	10.481	35.302	$3.759	$39.40	07/12/11	16	(0.20)	-5.05%	23.13	GV
36	104912	183	6.033	30.333	$3.799	$22.92	07/19/11	23	(0.16)	-4.04%	7.96	GV
37	105302	390	10.324	37.776	$4.049	$41.80	07/20/11	8	0.29	7.71%	48.75	Elko, NV
38	105668	366	10.008	36.571	$3.669	$36.72	07/21/11	1	(0.38)	-9.39%	366.00	Idaho Falls, ID
39	106027	359	9.565	37.533	$3.599	$34.42	07/26/11	5	(0.07)	-1.91%	71.80	Pocatello, ID
40	106396	369	10.445	35.328	$3.839	$40.10	07/26/11	1	0.24	6.67%	369.00	Battle Mountain
41	106783	387	11.083	34.918	$3.799	$42.10	08/03/11	7	(0.04)	-1.04%	55.29	GV
42	107092	309	12.003	25.744	$3.899	$46.80	09/07/11	35	0.10	2.63%	8.83	GV

Are we there yet? Figuring distance, time, and speed

When you travel in an automobile, you can't help but ponder life's deep questions. How long have I been driving? How much more time will my drive take? How far have I traveled? How many miles are left to go? What's my average speed? It's handy to be able to compute distance, time, and speed, because the answers take some of the mystery out of making a trip. (Doing the calculations helps fight boredom, too.)

Distance, time, and speed (formally called *velocity*) combine to make a great formula. When you know two of the items, you can solve for the third. Here are the three formulas:

$$\text{distance} = \text{time} \times \text{speed}$$

$$\text{speed} = \frac{\text{distance}}{\text{time}}$$

$$\text{time} = \frac{\text{distance}}{\text{speed}}$$

Estimating arrival time

Estimating when you'll arrive at your destination is an easy calculation using the formula for time. You need to know the miles left to drive (info you can get from highway signs) and your speed (just glance down at your speedometer). Say that you have 75 miles to go and are driving at 60 mph. With these two bits of information in hand, apply the time formula:

$$\text{time} = \frac{\text{distance}}{\text{speed}}$$

$$\text{time} = \frac{75}{60}$$

$$\text{time} = 1.25$$

You should arrive in 1.25 hours.

Don't want to do the math yourself? If you have a GPS, it'll do the job for you.

GPS

A global positioning system (GPS) navigation device is filled with amazing information. After you enter a destination, the GPS tells you the estimated time of arrival (ETA), remaining hours and minutes until arrival, remaining distance, and your current speed. As a bonus, you also get an irritating voice that says, "Return to the nearest road."

Actually, your GPS is more accurate than your dashboard instruments, in particular the speedometer. Speedometers are subject to mechanical errors, and you can't calibrate them. By contrast, your GPS is reaching out to at least three satellites at any given moment, and the data is recomputed about every 50 feet of travel.

Quick math for distance traveled and distance left to drive

If you can subtract (and you can), you'll have no problem calculating how far you've traveled or how far you have left to drive.

To see how far you've traveled, look at your trip *odometer*, an instrument that shows distance traveled. Some cars have as many as three odometers: the regular one, Trip 1, and Trip 2. You can switch between them and can easily reset Trip 1 and Trip 2 to 0.

If you have only one odometer, just subtract its current reading (your current location on the road) from the reading you took on your departure.

Distance traveled = current location reading – departure reading

Distance traveled = 23,331.5 – 23,130.0

Distance traveled = 201.5

Keep in mind, however, that real-life math often requires no math at all. If you have a trip odometer, you can find the distance traveled just by resetting the odometer to 0 at the start of your trip and reading the mileage at any time during the trip.

You can use Trip 1 for measuring the distance covered on an out-of-town day trip, while using Trip 2 to accumulate how many miles you've driven on the current tank of gas.

Speed-o-meter

The *speedometer* is an instrument that tells you how fast you're going. It was invented in 1888, and people have been speeding ever since. It shows speed in miles per hour (mph) and kilometers per hour (kph). Most speedometers have analog displays (like an older wristwatch), but some have digital displays.

Driving at or below the speed limit should keep you free from speeding citations (a real money-saver). And, generally speaking, driving faster takes more gasoline (or diesel fuel) and lowers your mileage.

Want to see how far you have left to drive? Here's another non-math real-life math solution: Use the highway mileage signs.

Estimating average speed

Your speedometer shows your speed at any given moment, but a drive might include stop-and-go traffic in town as well has fairly smooth driving on the highway. Calculate your average speed by dividing the total distance you've driven (from the odometer) by the time you've been traveling (from the car's clock or your watch). For example, if you've driven 50 miles in 2 hours — 50 ÷ 2 = 25 — your average speed is 25 miles per hour.

Making sense of the mechanic's bill

Your automobile mechanic charges for *materials* and *time* (or *labor*). Your bill reflects parts used and labor hours expended. But if you want to know the cost before you commit to a repair, simply ask for a written estimate (in which the shop will calculate the approximate cost for you).

If you don't want to take your car in for an estimate, do some Internet research to find the cost of the parts you need and the industry standard time required to complete the job. Then just ask your mechanic what the shop's hourly labor rate is. Now that you know these things, you can use math to create your own estimate.

 Self-estimating works when you know just what you want to do (for example, replace a fuel pump). However, if your car has an unknown problem (for example, "It makes a funny noise when I turn the wheel to the left"), you should take it into the shop for a diagnosis.

The formula for time and materials is very simple. You just add up the cost for the labor (which you calculate by multiplying the number of hours worked by the shop's hourly labor rate), the cost of the parts, and any applicable tax. Here's an example of the formula with values plugged in:

$$\text{total charges} = (\text{hours} \times \text{hourly rate}) + \text{parts} + \text{tax}$$

$$\text{total charges} = (4.17 \times \$89.928) + \$390.40 + \$30.74$$

$$\text{total charges} = \$375.00 + \$390.40 + \$30.74$$

$$\text{total charges} = \$796.14$$

In this example, hours are figured to two decimal places (hundredths of an hour). The labor rate is figured to three decimal places. The sample calculation is exactly what it cost me to replace a timing belt on my 2005 Honda Civic.

The charges are based on the cost of parts and the cost of labor. Figure 9-1 shows the detail from the automobile repair bill.

Part Description / Number	Qty	Sale	Extended	Labor Description	Hours	Extended
Timing Belt Kit				Cust wants timing belt replaced	N/A	N/C
2523120	1.00	218.90	218.90			
Water Pump New				TIMING BELT - Remove & Replace -	4.17	375.00
41115	1.00	68.15	68.15	1.7L Eng - [Includes: Adjust Valves.]		
Serp Belt				Replace timing belt and all timing belt		
25-060388	1.00	23.67	23.67	pullies, water pump and drive belts. Refill		
Serp Belt				system with new coolant and test drive.		
25-040398	1.00	17.95	17.95			
Honda Long Life Antifreeze						
OL999-9011	1.00	25.18	25.18			
Lower Timing Belt Cover						
1811-PLC-000	1.00	36.55	36.55			

Illustration by Wiley, Composition Services Graphics

Figure 9-1: Details on an automobile repair bill.

When you take your car in to be serviced, you'll likely be given a less-detailed written estimate. When you get the final bill, you'll see the details.

As Figure 9-1 shows, you multiply the quantity ("Qty") by the price ("Sale") to get the full cost of a part ("Extended"). Labor charges, on the other hand, can be a little bit baffling. To get this amount, the mechanic looks up the repair in a repair labor guide ("the book"), which is a list of standard rates for the type of job to be done. The book lists the time it generally takes to perform a given repair (a timing belt job is 4.17 hours, for example, as is the case in this figure). Some guides also show "regular flat rate" and "severe-condition flat rate." The "severe-condition flat rate" applies when particular factors (like rust and corrosion) are present that necessitate more time and more tools to do the repair.

Dining Out

People everywhere love to dine out, whether they go to drive-ins, cafés, bistros, taverns, fast food drive-thrus, or five-star restaurants. Chances are you know what kind of cuisine or experience you're in the mood for. So whereas comparing prices is a key part of making other buying decisions, it's not so much a concern when you're choosing where to eat out. Instead, the math you're most likely to use at a restaurant involves figuring out how much to tip and how to split a bill when you're dining with others.

Calculating the tip

Although tipping doesn't exist in some countries, in the United States, your restaurant server expects a tip. Breakfast, lunch, or dinner, you always have to tip. The customary amount is between 15 and 20 percent of the bill. Sometimes you tip a little more; sometimes you tip a little less.

The easiest way to calculate a 15 percent tip is to follow these steps:

1. **Divide the check by 10, which gives you 10 percent of the total.**

2. **Take half of the 10 percent, giving you 5 percent.**

3. **Add both values together, and that's your 15 percent tip.**

Here's an example for a $16.00 meal:

$$\text{tip} = (\text{check price} \times 0.10) + (\text{check price} \times 0.05)$$
$$\text{tip} = (\$16.00 \times 0.10) + (\$16.00 \times 0.05)$$
$$\text{tip} = \$1.60 + \$0.80$$
$$\text{tip} = \$2.40$$

That's your tip. If you like, round the $2.40 up to $3.00.

Calculating a 20-percent tip is even easier. Simply divide the total by 10 and then double the result. Here's the math: $16.00 ÷ 10 = $1.60 × 2 = $3.20. Your tip is $3.20.

You tip in other places, too, and the amount of the expected tip varies by type of service and by region. Here's a quick guide to what you should tip for different services:

Service/Person	*Customary Tip*
Hair stylist	15–20 percent
Shampoo person	$2.00 ($5.00 in Beverly Hills)
Manicurist	15 percent
Hotel bellhop	$1.00 per bag
Hotel maid	$5.00–$10.00 per stay
Skycap at airport	$1.00 per bag
Bartender	$1.00 per round of drinks
Coatroom attendant	$1.00
Valet	$1.00 at the mall, $2.00 at a nice hotel

Splitting hairs and bills

You've probably gone out with a group of friends or the gang from the office. Did you ever get skinned when the bill came? Though they mean no harm, some of your friends or co-workers may miscalculate what they owe, forget drinks they ordered, and get amnesia about tipping, leaving you holding the bag. The old rule of thumb for work lunches used to be, "Don't be the last one to leave the table."

No more. When your group is settling up, pull out your smartphone, the one with the great calculator, and then calculate each person's food, tax, and tip charges. Here's the calculation:

$$total = (food + drink) + tax + tip$$

Follow these steps:

1. **Figure out the total food and drink cost for each diner.**

   Say that your friend Wanda ordered a cheeseburger ($6.00) and coffee ($2.00).

   $6.00 + $2.00 = $8.00

   Food and drink amounts to $8.00.

2. **Figure the amount of tax for this diner and then combine the tax with the food and drink cost.**

   Say that sales tax in your area is 9 percent.

   $$tax = (food + drink) \times tax\ rate$$
   $$tax = \$8.00 \times 0.09$$
   $$tax = \$0.72$$

Wanda's share of the sales tax is $0.72. Combine that with the food and drink cost and she owes $8.72 before you calculate the tip.

Some states charge higher than normal tax rates (as high as 20 percent total) for mixed drinks. In Arkansas, the extra tax is called a "supplemental mixed drink tax." Keep this in mind as you calculate the tax.

3. Calculate this diner's tip amount.

Suppose that your group has agreed that a 15 percent tip is fair.

tip = (food, drinks, and tax) × tip percentage

tip = $8.72 × 0.15

tip = $1.31

Wanda's share of the tip should be $1.31. Now return to the original full calculation:

total = (food + drink) + tax + tip

total = $8.00 + $0.72 + $1.31

total = $10.03

Wanda's share of the bill is $10.03.

4. Repeat this process for everyone in your party.

If you do, you'll have enough money to pay the check.

For large groups (usually eight or more people, although some restaurants consider a group of six to be "large"), restaurants love to charge *mandatory tips*. In the Olden Days, the amount was 15 percent, but it has risen to 18 percent and sometimes 20 percent. The law says that a tip is mandatory when it's written on menus, in brochures, or in ads. And, if anybody in your group asks you, mandatory tips are subject to sales tax.

Taking a Vacation: To Drive or to Fly?

Some people would say that visiting the Walt Disney World Resort (informally known as Disney World) or similar theme parks is a dream vacation. Others prefer quiet cabins in the woods or cottages on a beachfront. Some like travelling to places where they can take in shows or museums.

Whatever your dream vacation is, you can calculate the full cost pretty easily, simply by adding up the travel expenses, the lodging expenses, any necessary ticket purchases, food, and other things, such as souvenirs and excursion costs. Figure 9-2 shows expenses anticipated by a family of four planning a trip to Walt Disney World.

THE WHOLE VACATION	
Item	Cost
Airfare	$1,800.00
Hotel	$1,204.00
Disney World	$952.00
Rental Car	$329.00
Food	$1,120.00
Souvenirs	$300.00
TOTAL	$5,705.00

Illustration by Wiley, Composition Services Graphics

Figure 9-2: Costs associated with a Disney World vacation.

The quandary that many families face is whether to drive or fly. To decide, you need to figure up the costs associated with each mode of transportation and then compare them. In the following sections, I help you identify the different expenses and explain how to do the calculations that can help you decide.

Leaving on a jet plane

Traveling by air has obvious advantages, the most significant being how quickly you arrive at your destination. But is it always the best choice, especially when you're flying with others (like your spouse and kids)?

The first thing to consider is the cost of the airfare. How much does a round-trip ticket cost? If you buy at the last minute, you'll pay full fare, but if you buy well in advance on the Internet, you'll save some big bucks. Multiply the cost of a single ticket by the number of people traveling with you. Say that your family is made up of you, your spouse, and your two children, ages 8 and 10. (By the way, both of your children are "adults" on an airline.) You multiply the airfare by 4.

 To determine airfare costs, simply go to the airline's site to see what flights are available and how much the fare is. You can exit the site without buying anything, if you're just investigating.

When you travel by air, however, the cost of the plane tickets isn't the only travel expense you'll incur. Here are other costs you need to figure in your calculations:

- ✔ **Long-term parking:** To figure parking, multiply the daily rate by the number of days you'll be gone. Of course, you can ignore this cost if your kind brother offers to drive your family to the airport.

- ✔ **Car rental or transportation fees:** Unless you are going to a resort that you don't plan to leave, chances are you'll rent a car at your destination or take public transportation or taxis to get to various places you want to go.

After you have the amounts for the expenses you'll incur if you fly, add them together. This is the amount you'll compare to the costs associated with driving.

As an example, say you book airfare well in advance of your vacation. The fare is $275.00 to get there and $175.00 to get back. Together that's $450.00 per traveler. Since four people are in your party, the airfare would total $1,800.00.

As a bonus, assume that your family will be dropped off and picked up at the airport (no cost!) and that you'll take a free shuttle from your destination airport to the resort.

Driving: The daring alternative

You didn't grow up to become a cost accountant, but sometimes you have to act like one. You need to do the calculations necessary to see whether you save money by driving to your destination. Fortunately, the math is pretty simple: addition, subtraction, multiplication, and division.

To calculate roughly how much driving will cost, you first need to determine the distance, which will give you an idea of how long you'll be on the road. Fortunately, you don't even have to make assumptions or guess. You can use an Internet map to find exactly how many miles you'll drive to reach your destination.

Say you're diving from Sacramento (near where I live) and heading to Disney World. You'll drive 2,890 miles. That's 45 hours of driving, at an average speed of 65 mph. If you can stand driving 8 hours per day, you'll be on the road for a little more than 5 days (45 ÷ 8 = 5 with a remainder of 5). You drive into Orlando on the 6th day.

After you know how many miles you're driving and how many days you'll be on the road, you can do the other calculations:

- ✔ **Gasoline:** To determine how much gas you'll use, divide the number of miles by the miles per gallon your car gets. If you can get 30 miles per gallon in gas mileage, you'll use 93.33 gallons getting there (2,890 ÷ 30 = 96.33). Because you also have to drive back, double that to get 192.67 gallons. (The gas you use driving around Orlando is the same as you'd use in a rental car, so it doesn't matter for this calculation.) But to make the math easy, call gas consumption 200 gallons.

  To figure how much you'll pay for those 200 gallons, multiply that by the price of a gallon. If you assume gasoline averages $4.00 per gallon over the whole trip, you'll pay about $800.00 for the gas.

- ✔ **Hotel:** You have to figure the hotel costs for every night you spend on the road. The good news is that hotels rates are generally cheaper along the route than they are in popular destinations. Maybe you can find a place to stay for about $90.00 night. Multiply that amount by the number of nights you'll be traveling. In the Disney example, you'll be staying at hotels for 5 nights and driving into Orlando on the 6th day. Your on-the-road hotel costs are $450.00 ($90.00 × 5 = $450.00).

- ✔ **Meals:** For each day travelling, figure in the cost for all the meals you'll eat en route. In the Sacramento-to-Disney World example, you're on the road for 5 days, you have 4 people in your party, and you eat 3 times a day. Multiply to get the number of meals you'll buy while traveling: 5 × 4 × 3 = 60 meals. Estimate an average meal cost per person per meal — say it's $5.00 — and multiply that by the number of meals: 60 × $5.00 = $300.00 for food.

Now, add all the items together: $450.00 + $800.00 + $300.00 = $1,550.00. If you compare travel by car ($1,550.00) to travel by air ($1,800.00) you save a little bit of money ($250.00) by driving.

There are hidden costs (and maybe benefits) in driving and you should be aware of them. You will be spending more than 10 travel days in a car with your family. The experience could bring you all closer or make you crazy. Also, consider *opportunity cost*, the value of the next best use of your time. You, your spouse, and your children will lose 10 days on the road. Maybe there are better things to do with your time than driving 5,780 miles.

Gambling: Money You Take to Las Vegas Stays in Las Vegas

Ah, the lure of the tables! Las Vegas, Atlantic City, and the Mississippi River are big destination resorts where, rumor has it, you can play casino games. Native American gaming is also wildly popular, and there are 400 of these establishments. Besides big casinos, many other forms of gambling exist. Some of the most popular are lotteries (in the United States, 43 states and the District of Columbia have lotteries), office sports pools, and bingo.

If gambling appeals to you, you'd probably prefer winning to losing. That means that you should know some of the math involved and be able to calculate one or two important numbers.

Understanding odds, bets, and payouts

Every casino game consists of placing a bet (sometimes called *making a wager*). Betting applies to non-casino games, too, such as card games and parimutuel betting ("playing the ponies").To state the obvious, you bet money that a favorable outcome will occur. If it does, you get a payout.

Knowing your odds

Odds are the ratio of an unfavorable outcome to a favorable outcome. Odds are essentially the same as probability, but expressed differently (refer to Chapter 3 more info and some math examples related to probability). In rolling a die, the *probability* of rolling a 1 is 1 chance in 6, or 1/6. There's 1 favorable outcome and 5 unfavorable outcomes. The *odds* are expressed as 5:1, and you'd say it as "5 to 1 against."

There are odds for everything from a roll of the dice in craps to drawing to an inside straight in poker (which is known as a fool's bet or sucker bet, because the odds are 47:4 against). If you want to be an informed gambler, know the odds associated with the game you're playing. Use the Internet and books to learn particulars of your favorite games.

Understanding types of bets

There are many classes of bets. Concern yourself with the two major ones:

- ✔ **Fair odds bet:** A *fair odds* bet pays off in exact proportion to the odds. If two people make a bet with each other, without the services of a bookmaker or casino, the return is equal to the risk. For example, if two people bet $1.00 that something will happen, the winner gets his or her $1.00 dollar back plus the other person's $1.00. The loser loses $1.00. An office sports pool usually works this way.

- ✔ **Viggorish bet:** A bet that includes an allowance for the person or business conducting a game (a bookmaker or a casino) isn't a fair odds bet. If it were, the "house" would go out of business. I explain how a viggorish bet works in the next section.

Knowing payouts and house edge

Gambling facilities aren't charities; they need to make money. Almost all bets include a fee of some sort.

The *viggorish* (also known as the *vig*, the *juice*, the *cut*, the *rake*, or the *take*) is the amount charged by a bookie or casino for giving you the privilege of gambling. The word comes from the Yiddish, probably from Russian, and means "winnings" (but not for you).

A common term is *house edge* (also known as *house advantage*). It's the percentage the house takes in on various bets. This term comes up again and again in assessing the best and worst bets. In some cases, the house edge is easy to calculate, but many bets require the work of a mathematician. (But don't worry; there are many sources that tell you the precise house edge.)

The closer to 0 percent a house edge is for a bet, the better the bet is for you. In general, any bet with a house edge of 2 percent or less (as in some craps bets, for example) is a good one.

Playing the most popular games

Should you visit casinos, you'll find (and maybe play) the most popular casino games. The only rules you need to follow are to pick your favorite games, know how to play them, and use math to help you improve your odds of winning.

The house edge and roulette

Some formulas for calculating the house edge are pretty simple. Take roulette. The roulette wheel has numbers from 1 to 36, plus 0 and 00, giving 38 possible places where the ball may fall. The payout is 35 times the bet (35:1). The difference between 38 (true odds) and 35 (payout) is important. In roulette, if you bet on a single number, you have a probability of 1/38 of winning and a probability of 37/38 of losing.

A common bet is one on "red or black." The payout is 1:1, so if you win, you get your bet back plus and equal amount in winnings. The table has 18 black numbers and 18 red numbers. The wheel has 18 black and 20 non-black numbers (which also means that it has 18 red and 20 non-red numbers).

To find the house edge, subtract the probability of an unfavorable outcome (20/38 in this case) from the probability of a favorable outcome (18/38 in this case), and multiply the result by 100. To make a long story short, the house edge for roulette is 5.26 percent, making a roulette bet a fairly bad bet. The number 5.26 applies to every possible bet on the table. Can a casino make money on the house edge of "just" 5.26 percent? You'd better believe it!

Here's a tip: If you're dying to play roulette, go to Europe. European roulette wheels have only a 0; there's no 00. As a result, the house edge drops to 2.70 percent.

Slots

The *slot machine* (also known as the *one-armed bandit*) is enormously popular — the most popular casino game of all. Casinos are filled with hundreds or thousands of slot machines.

Slot machines are a poor form of gambling, because people lose a good deal more than they win. Slot machines are typically programmed to pay out 82–98 percent of the money put in.

On the road to Reno (I live nearby), billboards advertise "loose" slots, which means the casino is promising bigger payouts. Local casinos put the increasing payout amount of progressive slots on electronic signs on freeway billboards! (A *progressive jackpot* is a huge amount — $250,000, for example — that increases with every play of a bunch of "linked" machines.) In 2006, a man won $21 million in a Nevada Megabucks jackpot. (You can watch the current Megabucks in Nevada payout amount climb at http:// www.megajackpots.com/games/megabucks-in-nevada. aspx.)

Mother knows best

For reasons I cannot explain, my mother never passed by a slot machine that didn't like her. She once showed me the results of a little trip to Las Vegas — eighteen $100 bills, all won playing quarter slots. She went to Vegas so often that the casinos thought she was a high roller and sent her coupons for free rooms, free meals, and free satin jackets.

Once, while my father was just checking in to the hotel, she put three quarters in a machine and immediately hit a $250.00 jackpot — before the ink was dry on the registration form. Her advice: Always play three coins, not two or one.

You hear a lot of lore about slots. One story is that casinos make them looser early in the week to attract more customers on off-days. Another story is that the loose machines are located near the main entrance, so the noise from jackpots attracts customers from the street. In the past, the casino made a slot looser by changing the colorful belts on the reels; today, the casino just reprograms the machine.

Blackjack

Customers like blackjack (sometimes called "21") because it's easy to play. Casinos like blackjack because customers lose so much.

With blackjack, you see the two cards you were dealt. You see one of the two cards the dealer dealt himself or herself. The object is to take more cards, as needed, to bring your hand up to 21 points without going over 21 points (*busting*).

To judge how close you are to 21, you add up the values of the cards. Each card is worth the number you see on it (its *pip value*). Jacks, queens, and kings have a value of 10. You can treat an ace as either 1 point or 11 points.

Based on your cards, the dealer's card, and some underlying probabilities, you decide whether to *hit* (take more cards), *stand pat* (take no more cards), *double down* (increase your bet and take only one more card), or *split* (divide a pair into two hands and make an additional bet).

✔ **When to stand:** Stand when you're holding 17–20 points, no matter what card the dealer is showing (the *upcard*).

✔ **When to hit:** Hit when your cards total 5–8 points, no matter what upcard the dealer is showing.

✔ **When to split:** Split a pair of aces or 8s, no matter what upcard the dealer is showing.

✔ **When to double down:** Double down when you're holding 10 or 11 points, except when the dealer's upcard is a 10 or an ace.

Broad strategies for winning at blackjack exist, too. Here's a list of strategies you can use to improve your odds and avoid bad bets:

✔ **Following the basic strategy:** The most important technique for winning at blackjack is to follow what's known as basic strategy, which says that, in playing any hand, you make the best decision possible (based on the work of professional mathematicians). Doing so can lower the house edge to less than 1 percent. You can visit http://en.wikipedia.org/wiki/Blackjack#Blackjack_strategy to see 230 distinct scenarios for decisions.

✔ **Counting cards:** With card counting, you track the relationship between high-value cards (good for the player) and low-value cards (good for the dealer). If you learn how to count cards, you could get a 1–2 percent edge *over* the casino. To become a "counter," prepare to spend maybe 150 hours practicing the technique, plus time learning how not to be spotted.

Very broadly, your betting strategy is to increase your bets when the deck is "running in your favor." Decrease your bets when the deck is "running against you." Don't be obvious, such as switching from a $5.00 bet to a $100.00 bet, because it draws the pit boss's attention.

Counting isn't illegal, not by any means. But casinos don't like it, and if you're spotted, they may permanently bar you, and they may alert other casinos. For the movie version of card counting, rent the film *Rain Man,* starring Dustin Hoffman and Tom Cruise.

✔ **Avoiding the hunch play:** In the *hunch play,* you deviate from basic strategy. Guessing at the outcome (and being ignorant of how to play the game) costs most people a lot of money.

✔ **Avoiding the insurance bet:** An "insurance" bet is a side bet you make when the dealer's showing an ace. It's a bad idea. The house edge is about 8 percent, so forget about it.

Craps

Craps is a dice game, prominent in the movies and Broadway musicals. It's based on the outcome of one or more rolls of a pair of dice. The one person throwing the dice (the *shooter*) bets on the outcome, but so does everyone else around the craps table.

Very briefly, here are the rules for craps: The shooter's first roll is the *comeout roll*. If the comeout roll is a 2 (*snake eyes*), a 3 (*craps*), or a 12 (*boxcars*), the shooter loses. If the comeout roll is a 7 or 11 (a *natural*), the shooter wins. Any other roll (4, 5, 6, 8, 9, or 10) establishes the shooter's "point." The shooter has to roll dice — again and again, if necessary — to either make the point or lose. The shooter makes the point in order to win, but if he or she rolls a 7 before making the point, he or she loses.

The shooter and everyone at the table place bets before the shooter rolls. When the shooter wins, everyone betting with him or her (a *passline bet*) wins. If the shooter loses, everyone betting against him or her (a *don't pass* bet) wins. There are other miscellaneous bets, most of them poor choices.

These bets are the best ones in craps:

- ✔ **The passline bet:** This is a bet that the shooter will make his or her point. The bet has a low house edge of 1.41 percent. Most players make this bet.

- ✔ **The don't pass bet:** This is a bet that the shooter won't make his or her point. The house edge is 1.14 percent.

- ✔ **The odds bet:** This is the *only* bet in the casino that doesn't have a house edge, meaning *it's paid off at true odds*. First, you place a bet on the pass line. If the shooter establishes a point of 4, 5, 6, 8, 9, or 10 on the comeout roll, you put an additional bet behind your first bet and ask for "odds." If the point is 4 or 10, the bet pays 2 to 1; if the point is 5 or 9, it pays 3 to 2; if the point is 6 or 8, it pays 6 to 5. Of course, the bet pays only if the shooter makes the point.

All the other bets in craps are bad bets. For example, *proposition bets* are bets on the "hard way" (4, 6, 8, or 10, with both dice having the same value). So are one-roll bets, including 2 (snake eyes), 3 (ace-deuce, or craps), 11 (yo-leven), and 12 (boxcars). The house edge can be as high as 16.7 percent, depending on the bet.

Poker

Poker is a card game where player skill is the biggest factor in determining the winner. Among other things (such as how to bet and how to bluff), the best players know their odds cold. For example, they know that in a game of five-card draw, the deck has 1,098,240 pairs (42.3 percent probability of getting one) and only 36 straight flushes (0.00139 percent probability of getting one).

In poker, players compete against each other and not against the house. So there's no house edge, except for a small amount the casino takes out of each pot.

Video poker is also a game of skill. The house edge is small: 0.1 to 1.4 percent. The payoff schedule may actually pay back *more* than 100 percent if you play perfectly.

The worst casino bets

The worst casino bets to make are those where the house edge is high. Here are three outrageously bad ones:

- **Roulette:** Bets on the American roulette wheel (with both a 0 and a 00) have a house edge of 5.26 percent. (For more on roulette, go to the sidebar "The house edge and roulette.")

- **Keno:** Keno is a game similar to the lottery, except that draws are held every few minutes. Bored diners in the casino coffee shop fill out forms with possible winning numbers and give them to a "runner" while they're waiting for food. Payoffs are low, and the house edge can be 25 percent or higher.

- **Big Six:** The Big Six wheel (Wheel of Fortune) is a large vertical wheel. The dealer gives it a spin, and it turns until a rubber pointer stops it at the winning number. The house edge is 11–14 percent, depending on the bet.

Part III

Math to Manage Your Personal Finances

In this part . . .

Personal finance may not be the easiest task in your life, but it's definitely one of the most important. The chapters in this part make investment, insurance, and business easier. Here you can find information on how to apply real-life math in the areas of banking, credit, investments, and insurance. Because these fields have a language all their own, I explain what the specialized terms and concepts mean.

As a bonus, I include a chapter on taxes because you can't manage your personal finances without knowing a bit about what goes to the IRS and your state treasurer.

Chapter 10

Budgets, Bank Accounts, Credit Cards, and More

● ●

In This Chapter

▶ Getting your budget and checkbook in order

▶ Understanding how interest and amortization affect your mortgage and other loans

▶ Avoiding credit card trouble

▶ Sifting through savings and other accounts

● ●

Do you ever worry about money? If so, you're a member of a very large club. Whereas the term *finance* used to mainly mean *high finance* — the province of giant banks, large corporations, and the government — now finance is everybody's concern. The reason is simple: Income used to be more reliable, expenses were lower, and ways to borrow were simpler. The world of *personal finance* has grown a lot more complex. Fortunately, your real-life math skills can help you get through the mire.

The great thing is that you don't need to be a genius to do personal finance. In this chapter, I cover budgets, loans, bank accounts, and credit — all of which are part of modern life. The math related to these topics centers around principal and interest, and it's is very much the same whether you're dealing with loans or savings. The rest of the math is simple addition and subtraction.

Beginning with a Budget

Broadly, a *budget* is a summary of income and expenses that you can use to manage money. A budget serves two purposes. First, knowing where your money comes from and goes to gives you increased control over it right now. Second, you can use a budget to plan for the future.

Identifying what's in a budget

A budget shows income and expenses from all sources. For most people, *income* refers to the salary they receive from their employers, but it may also include alimony (spousal support), interest, or Social Security benefits.

Expenses are everything you spend money on. A few big items (such as house payment, car payment, insurance, and groceries) account for most people's monthly spending, but the little things can add up, too. Figure 10-1 shows a very simple budget.

Category	Monthly Amount	
Income		
Salary	$2,600.00	$2,600.00
Expenses		
Rent	$900.00	
Car payment	$360.00	
Auto insurance	$100.00	
Food	$320.00	$1,680.00
Net		$920.00

Illustration by Wiley, Composition Services Graphics

Figure 10-1: A simple budget showing basic income and expenses.

The budget in the figure is a good start, but it's pretty incomplete. It shows income from one source, salary. The only expenses listed are rent, car payment, auto insurance, and groceries. It's a rosy picture — but it can be a bit misleading because it's not complete. Most budgets are more complex.

Using your math skills to make a budget

When you create a budget, you use addition, subtraction, multiplication, and division. (Head to Chapter 1 for a review of these basic operations.)

Although you can do a budget using paper and pencil, "automated" tools can make the process easier. They include online budget tools, smartphone apps, and money management software. I think the best tool for budgeting is the spreadsheet. You can easily update it, and it practically does the math for you.

If you don't want to pay for spreadsheet software, get OpenOfficeCalc, part of the Apache OpenOffice suite. It's owned by the Apache Software Foundation, a United States non-profit corporation. Go to `http://www.openoffice.org/` for more information.

To make a complete budget, follow these steps:

1. **List your income and expense items in categories.**

   List every income and expense item you can think of. Housing, car, insurance, and food are easy to remember, because they're usually the biggest expenses. But don't forget credit card payments, childcare, clothing, savings, medical expenses, and recreation.

   Even though certain expenses, such as doctor's visits come up infrequently, putting them in a monthly budget starts you thinking about saving to meet them.

2. **Record the dollar amount associated with each item.**

   You can note the cost of each weekly, monthly, or annual item, as Figure 10-2 shows. At this point, it doesn't matter whether an expense is annual (for example, a doctor's visit or vehicle registration) or weekly (for example, a music lesson). You'll convert everything to a monthly amount in the next step.

Category	Weekly	Annual	Monthly	Total
Income				
Net Salary	$480.00		$2,080.00	$2,080.00
Expenses				
Rent			$900.00	
Car payment			$360.00	
Auto insurance			$100.00	
Food			$320.00	
Physical exam (1 × $200)		$200.00	$17.00	
Dentist (2 × $80)		$160.00	$13.00	$1,710.00
Net				$370.00

Illustration by Wiley, Composition Services Graphics

Figure 10-2: Note categories and dollar amounts on your budget.

In this example, notice that salary becomes "Net Salary." A *net salary* is the gross salary (say $600.00 per week) with the payroll deductions subtracted. The result here is about $480.00 per week.

3. **Calculate as needed to convert weekly and annual items into monthly numbers.**

   Both calculations are easy. To convert a weekly salary into a monthly number, multiply it by 52 (the number of weeks in a year) and then divide it by 12 (the number of months in a year). Here's an example, using a weekly salary of $480.00:

   monthly salary $= ($ weekly salary $\times 52)/12$

   monthly salary $= (\$480.00 \times 52)/12$

   monthly salary $= \$24{,}960.00/12$

   monthly salary $= \$2{,}080$

   To convert an annual expense into a monthly expense, just divide by 12. Here's an example using an annual expense of $324.00:

   monthly amount $=$ annual amount$/12$ months

   monthly amount $= \$324/12$

   monthly amount $= \$27.00$

4. **Calculate net spendable income.**

   This amount is also called *disposable income* or *discretionary income*. To calculate it, simply subtract your total expenses from your total income.

   Your goal is to have a little money left over. If you don't have money left over, you're "under water" or "in the red," as many people are. See the next section for some ways you can get back in the black.

Applying budgeting principles

Your budget should help you see clearly what you make and what you spend. With that info at hand, it's time to take charge. Following are typical weak spots that a budget will show:

✔ **No money left:** If all your income is consumed by expenses, you're living close to the edge. One unexpected expense can put you under.

✔ **No savings:** A little money in savings can help you meet some disasters, such as a major automobile repair. No savings means no safety net.

✔ **No allowance for vacation:** All work and no play makes Jack (or Jill) a dull boy (or girl).

Each of these scenarios means that you need to make some changes. Fortunately, there's hope! Change isn't always easy, but here are some things you can possibly do to get your budget on firmer footing:

✔ **Make more money.** Increasing your income isn't easy, but it's possible. Maybe you can make more money by taking on a second job or selling items online. If increasing what you bring in is impossible, don't despair. There are other ways to solve this problem.

✔ **Reduce expenses.** See where (and if) you can lower expenses. You're probably stuck with some payments, but maybe you can economize on food or entertainment. See Chapter 5 for information about reducing your shopping expenses.

✔ **Live within the budget.** At the very least, don't spend more than you already do. Technically, this is called a *budget constraint.*

✔ **Pay down credit cards.** Credit cards suck more blood than Count Dracula or anyone in the cast of *Twilight.* When the cards are paid off, don't load them up with debt again.

✔ **Stop spending on things you don't care about.** Take a hard look at things that aren't really important to you. Do you really need 250 channels on your cable or satellite TV? Do you really lust for fast food? If not, reduce those expenses or let them go altogether.

✔ **Set goals.** Say you want a vacation more than a double bacon cheeseburger. After you let go of spending on fast food, set a vacation goal. Any vacation saving, even as little as $10.00 per week (about $40.00 per month), is a good start toward meeting your goal.

Balancing Your Checkbook

Balancing a checkbook, as you very likely know, means keeping the balance in your check register in agreement with your bank statement.

The system isn't perfect, because invariably you make new deposits and write new checks while the bank is preparing and mailing your monthly statement. The end result is that you have a few items in your check register that don't appear on the statement, but you still have to account for them.

Because the task can get confusing if you aren't careful, some people avoid checkbook balancing. Just the same, like eating your vegetables, balancing is good for you. You've got to do it, or you won't know how much money you really have in your bank account.

Even though balancing a checkbook requires only simple addition and subtraction skills, the process may seem difficult until you get used to it. Here's a simple checkbook balancing method that really works:

1. **Find the date to reconcile to.**

   On the bank statement, find the date of the last transaction, whether it's a check clearing, a deposit, or a charge. This date is the one you reconcile to.

2. **Find the same (or closest) date in your register.**

   Draw a line in your register under the date closest to the statement closing date. There's no need to worry about deposits you've made or checks you've written after that date.

3. **Write both balances — the statement balance and register balance — on the back of your statement (see Figure 10-3).**

4. **In your register, check off cleared items and draw a little circle next to uncleared items.**

   Cleared items are checks, deposits, and ATM transactions that the bank shows it processed.

5. **Under the bank's ending balance, write any uncleared items from your register; then subtract the uncleared checks (very likely) from the balance and add uncleared deposits (not so likely) to the balance.**

Statement		
10/31	Ending balance	$477.21
Register		
10/29	Ending balance	$277.45

Illustration by Wiley, Composition Services Graphics

Figure 10-3: Write down the balances on the back of your statement.

6. Under your register balance, write any unrecorded items from the bank statement.

Typically, this includes monthly fees and interest, if you have a checking account that pays interest. It might also include a check you wrote or a deposit you made that you forgot to enter in your register.

In the example in Figure 10-4, you've recorded a credit card payment that you know about but that the bank hasn't processed yet. You've also recorded a monthly fee that the bank knows about, but that you haven't processed yet.

You're close, but the two balances don't agree! What's wrong?

7. If the balances don't agree, check your register for math errors.

In the example, you can see a difference of $0.24. So when you find an error (usually a math error or a check or deposit having been recorded incorrectly), record the adjusting item, as shown in Figure 10-5.

This is great! After adjusting for the error, your register balance and the bank's balance agree — to the penny!

8. Tidy things up.

Make sure your register has an entry for each adjusting item, including bank fees and/or interest and math/record-ing errors. It's a must, or you'll have problems when you reconcile next month. It's also nice to mark "AGREES" at the reconciling line.

 Useful symbols for your register are DEP (deposit), EFT (electronic funds transfer), ATM (ATM or debit card transaction), CHG (for monthly fees), and INT (interest).

Statement		
10/31	Ending balance	$467.21
#1024	Visa payment	-$200.00
		$267.21
Register		
10/29	Ending balance	$277.45
Charge	Monthly fee	-$10.00
		$267.45

Illustration by Wiley, Composition Services Graphics

Figure 10-4: Noting uncleared items.

Statement		
10/31	Ending balance	$467.21
#1024	Visa payment	-$200.00
		$267.21
Register		
10/29	Ending balance	$277.45
Charge	Monthly fee	-$10.00
Adj	Math error	-$0.24
		$267.21
	AGREES!	

Illustration by Wiley, Composition Services Graphics

Figure 10-5: Adjust for an errors in your register.

Homing in on Mortgage Math

In its simplest terms, a *mortgage* (called a *deed of trust* in about 21 states in the United States) is the loan you get to buy your home. If you make the payments, you end up owning the house. If you don't make the payments, the bank forecloses.

Your *first mortgage*, the *first trust deed*, is the most important home loan. There are other kinds of home loans, but they are *subordinate* to the first trust deed.

To put it simply, you pay a *down payment* (typically 20 percent of the home's selling price), and the bank lends you the rest of the price of the home. The amount loaned is the *principal*, and you pledge the house as *security* (a valuable asset the bank can take if you fail to pay off the loan). The loan accrues *interest* (the money the bank charges for the loan). You pay off the loan by paying back the principal and interest. After that, nothing's simple.

Having a PITI party

Your loan payment is made up of four parts: principal, interest, taxes, and insurance — PITI.

Principal

The amount you borrow is called the *principal*. Calculate principal as follows:

principal = selling price – down payment

Principal is also called the *loan amount*. Other costs are associated with closing the loan transaction. They are one-time-only fees; they're separate from the basic loan, but they can usually be added to the loan amount.

Closing costs include title search, title insurance, recording fees, broker/agent commissions, application fees, appraisal fees, termite inspection fees, and home warranties.

Interest

You pay interest on the mortgage loan. Charging interest is how banks make their money. Interest rates available for home loans can vary widely, and sometimes offerings seem to change daily.

In two popular loans, the 30-year fixed interest loan and the 15-year fixed interest loan, the interest rate is steady over the life of the loan (30 years and 15 years, respectively). A *fixed-rate mortgage* (FRM) is a "plain vanilla" mortgage, which is a pretty good idea.

An adjustable rate mortgage (ARM) — also called a *variable-rate mortgage* — is another option. This type of mortgage has an interest rate that can *float,* or change, going up when interest rates go up and going down when they fall. ARMs are usually tied to (*indexed to*) an index, such as the London Interbank Offered Rate (LIBOR) that reflects the cost of the lender's borrowing the money to lend you. ARM interest rates are usually lower than fixed rate mortgage interest rates, which is one of the things that make them seem attractive, but remember, they can adjust, and what comes down can go up.

Does a lower interest rate make a difference? You bet it does. The interest rate affects your monthly payment amount and the total you pay over the life of the loan. Table 10-1 compares four interest rates for a 30-year $180,000 loan.

Table 10-1	Interest Rates for a $180,000 Loan		
Interest Rate	*Monthly Payment*	*Total Interest*	*Total Cost*
3.0 percent	$758.89	$93,200.40	$273,200.40
3.5 percent	$808.28	$110,980.80	$290,980.80
4.0 percent	$859.35	$129,366.00	$309,366.00
4.5 percent	$912.03	$148,330.80	$328,330.80

Notice that each 0.5 percent increase in the interest rate sends your monthly payment up by about $50.00. Notice that each 0.5 percent increase sends the total cost of the loan up by about $18,000.00.

For math for real life, don't get hung up using formulas to determine loan figures yourself. Instead, use an online mortgage loan calculator, such as this one: `http://www.bankrate.com/calculators/mortgages/mortgage-calculator.aspx`.

To reduce monthly payments and the total cost of the loan, either shop around for a good interest rate or pay *points*. A *discount point* is a form of pre-paid interest. Some call it a bribe for the bank. Others call it a gamble that the money you pay for points will save you interest charges in the long run. Commonly, purchasing points is called a *buydown*.

Typically, each point you can buy reduces the interest rate by 1/8 percent (0.125 percent), and a point typically costs 1 percent of the loan amount. In general, the bank may offer you a chance to buy two points, resulting in a 0.25 percent drop in the interest rate. Figure 10-6 shows a very simple buydown comparison.

BUYDOWN COMPARISON			
Rate	Monthly Payment	Total Interest	Total Cost
4.50%	$912.03	$148,330.80	$328,330.80
4.25%	$885.49	$138,776.40	$318,776.40
DIFFERENCE	$26.54	$9,554.40	$9,554.40

Illustration by Wiley, Composition Services Graphics

Figure 10-6: A buydown comparison.

With a $180,000 loan, buying down from 4.50 percent to 4.25 percent costs you $3,600 (2 percent of $180,000). Is it worth it? Given that it reduces the monthly payment by $26.54 and reduces the total interest by $9,554.40, it seems like a pretty good deal.

A discount point isn't the same thing as a *loan origination point*. That's a scheme where you pay for some of the costs of the loan origination process.

Taxes

Property taxes are part of home ownership, and lenders aren't exactly filled with trust that you'll pay. To ensure you pay taxes on your property (which is really the bank's property until you pay your mortgage off), your loan payments may include an allowance for taxes.

In California, where I live, we typically use 1.25 percent of the purchase price as an estimate of annual property taxes. For a $200,000 home, calculate taxes as follows:

$$\text{tax} = \text{purchase price} \times 1.25 \text{ percent}$$

$$\text{tax} = \$200,000 \times 0.0125$$

$$\text{tax} = \$2,500.00$$

Insurance

Private mortgage insurance (PMI) is also known as *lenders' mortgage insurance*. PMI insures against your defaulting on the loan. The bank gets the benefit, and you pay the premium.

Here's how PMI works: If you default and the bank repossesses, the bank may not be able to sell the foreclosed home for a lot of money. PMI helps pay the bank back for its loss.

As I mention earlier, the bank isn't especially trusting. It's even less trusting if you don't have "skin in the game" — that is, if your down payment is low. A mere 10 percent down payment means the *loan to value ratio* (LTV) is 90 percent. That's a relatively high number. The LTV is important because it determines whether or not PMI is required. If your LTV is 80 percent or less, you may be able to forego private mortgage insurance altogether.

Here's how you calculate LTV. In this example, the cost of the home is $200,000 and the loan amount is $180,000 (which means you made a $20,000 down payment):

$$\text{LTV} = \frac{\text{loan amount}}{\text{purchase price}} \times 100$$

$$\text{LTV} = \frac{\$180,000}{\$200,000} \times 100$$

$$\text{LTV} = 0.80 \times 100$$

$$\text{LTV} = 80 \text{ percent}$$

Beware the low down payment! If you have an LTV of 90 or 95 percent, PMI may be as high as 1.15 percent of the loan amount per year. The way to get a better (read, lower) LTV is to make a higher down payment.

To see how much your annual PMI premium is, use this calculation:

$$\text{annual PMI premium} = \text{loan size} \times 1.15 \text{ percent}$$

$$\text{annual PMI premium} = \text{loan size} \times 0.0115$$

$$\text{annual PMI premium} = \$180,000.00 \times 0.0115$$

$$\text{annual PMI premium} = \$2,070.00$$

In this example, the home costs $200,000, with a 10 percent down payment of $20,000. The resulting LTV is 90 percent. The PMI premium for the year is $2,070, or about $172.50 per month!

PITI: Putting it all together

To see the approximate monthly bottom line of a mortgage payment, you add up the monthly principal, interest, taxes, and insurance.

PITI = principal + interest + taxes + insurance

PITI = $283.28 + $525.00 + $208.00 + $175.00

PITI = $1,191.28

This amount is an approximation, as it reflects only the first month in a 360-month loan. Over the course of the loan, the principle and interest vary slightly every month. See the next section for info about amortization.

Amortization: Paying down the loan

Amortization is the gradual paydown of a loan. The term comes from the Old French *amortir* — to reduce to the point of death. So you "kill off" your mortgage by paying it down.

Each monthly payment pays a bit of principal and a bit of interest. As the balance of the loan goes down, you pay a little less interest and a little more principal each month. To see what's happening, the lender gives you an *amortization schedule*. Figure 10-7 shows an amortization schedule for the last six months of a loan.

Month/Year	Payment	Principal Paid	Interest Paid	Total Interest	Balance
Feb 2042	$983.88	$963.23	$20.65	$154,144.88	$4,867.56
Mar 2042	$983.88	$966.64	$17.24	$154,162.12	$3,900.92
Apr 2042	$983.88	$970.06	$13.82	$154,175.94	$2,930.86
May 2042	$983.88	$973.50	$10.38	$154,186.32	$1,957.36
Jun 2042	$983.88	$976.95	$6.93	$154,193.25	$980.41
Jul 2042	$983.88	$980.41	$3.47	$154,196.72	$0.00

Illustration by Wiley, Composition Services Graphics

Figure 10-7: An amortization schedule.

Notice that monthly payments are fixed, but the amount of principal paid rises a little each month and the amount of interest paid decreases a little each month. At the beginning of the 30-year period, interest was the biggest component of the monthly payment.

How do you calculate amortization? Use an online calculator. Although you could do it by hand with the following formula for the payment (*p*), it's not practical:

$$p = \frac{P_0 \cdot r \cdot (1+r)^n}{(1+r)^n - 1}$$

In this formula P_0 is the initial principal, *r* is the monthly percentage rate (the annual percentage rate divided by 12), and *n* is the number of payments (typically 360 payments in 30 years). Too much math for my taste!

Reducing how much interest you pay

There are a few ways you can reduce the amount of interest you pay on your mortgage: Shorten the term from a 30-year fixed rate mortgage to a 15-year fixed mortgage, make one or more lump sum payments, or pay a little extra on principal each month:

✔ **Shortening the term:** If you have the income (and if the bank agrees), you may qualify for a shorter-term mortgage, such as a 15-year fixed. If you already have a 30-year fixed, you may be able to refinance to a 15-year fixed.

With a 15-year fixed, you pay more per month, but you pay off the loan in 15 years, saving yourself a bundle of interest in the process. Look at this comparison between a 30-year fixed and a 15-year fixed mortgage for a $200,000 loan. As you can see, you pay $524.55 more each month, but you save $77,451.36 in interest over the life of the loan. You also own your home sooner. Your best tool for mortgage comparisons is an online mortgage calculator. Visit http://www.bankrate.com/calculators/mortgages/mortgage-loan.asp.

30 VS. 15 YEAR COMPARISON				
Term	Rate	Monthly Payment	Total Interest	Total Cost
30 years	4.00%	$954.83	$143,739.01	$343,739.01
15 years	4.00%	$1,479.38	$66,287.65	$266,287.65
DIFFERENCE		−$524.55	$77,451.36	$77,451.36

Illustration by Wiley, Composition Services Graphics

✔ **Making lump sum payments:** *Lump sum payments* are the payment of a large amount of the principal at one time. Say, for example, you get a bonus each year and put your entire bonus toward the principal of your loan. Such payments typically don't shorten the term of the loan, but they do reduce the total interest paid.

✔ **Paying a little extra:** Pay a bit extra on principal each month. Increasing each monthly payment by a few hundred dollars (maybe just $100.00) helps bring the balance down more quickly, and that reduces the total interest paid.

A second mortgage or home equity line of credit

A *second mortgage* or *second deed of trust* is a second loan secured by your home. Second mortgages are subordinate to first mortgages. If you default on your second mortgage, the first mortgage gets paid off first.

You can also call a second mortgage a *home equity loan*. You would most likely "take out a second" to finance major expenses, such as college or big home improvements. Such a loan is typically a fixed-term loan.

In the United States, another form of home equity loan, the *home equity line of credit* (HELOC) is common. It's a useful, but somewhat risky, way to "pull money" out of your home. It's a revolving line of credit, so you can think of it as a variable-term loan. That means "it never ends." (I explain revolving credit in the section "Understanding how credit cards work.")

With a HELOC, you borrow only sums you need, up to a credit limit. The risk is that you will use the money to fund your lifestyle, not to meet major expenses. Should home prices go down, you may find that you're "under water." That is, what you owe on your house is greater than its value.

Using Smart Math for Other Big Purchases

Your home loan (or "first") is a biggie. So is a second mortgage or home equity line of credit. But they aren't the only kinds of useful loans. Two other common loans that typically involve large amounts are car loans and education loans.

Cruising around car loan math

An automobile loan is possibly the second biggest loan you have. With cars costing thousands of dollars (some as high as $32,000 to $60,000 or more), very few people have the ability to pay cash.

The total cost of a car loan is easy to see:

> total cost of loan = agreed-upon price – trade-in – down payment + tax + interest

The loan will be amortized over 36 to 72 months. In the "old days," 36 months was standard, but as cars got more expensive, the terms of auto loans increased to make the cars appear more affordable.

When you know the basics of the loan you want (total price, term, and interest rate), use an online car payment calculator to figure your monthly payments. For example, you could visit `http://www.bankrate.com/calculators/auto/auto-loan-calculator.aspx`.

Here are some things to keep in mind about auto loans:

- ✔ **If you find a dealer who advertises "We approve everyone," you will be probably be getting into a pretty bad loan.** Dealers who lend to higher-risk purchasers are willing to assume the risk of likely defaults because they make up the cost by charging high interest rates. In other words, the cost of the bad loans is paid by everyone else's loan.

- ✔ **A credit union is the best place to get a car loan.**

- ✔ **If you can, sell your old car privately.** You can get a better price than any dealer will give you on trade-in. Then you'll be able to put a larger down payment on the new car.

- ✔ **If possible, buy a car directly from the owner rather than a dealer.** Chances are, you'll be able to get a better deal because private sellers don't charge commissions and don't have to run a dealership at a profit. Although a private seller obviously wants to get a good price for the car, he or she also may be motivated by nonfinancial considerations, such as just getting rid of the vehicle to make room in the garage. In addition, because you have to save up the money for a private purchase, you may be able to avoid a loan altogether and will own the car the moment you buy it.

- ✔ **Learn all the fees before you buy from a dealer.** Expect to pay vehicle registration fees (since your car has to have license plates) and sales tax. Some places want to *tax the full purchase price of the new car* and ignore the cash value of trade-in.

Be wary of dealers who tax the full purchase price of the car and ignore the value of rebates (a practice many states permit), documentation ("doc") fees, and add in any fees labeled "dealer prep" or "shipping and handling." These last two fees are bogus fees.

Studying up on education loans

Some loans — specifically those that let you purchase things that end up worth more than the loan costs — are good investments. In that way, a student loan can be a great investment. These loans are great financial tools, designed to help students pay for college costs, such as tuition, books, and living expenses. In addition, student loans are relatively easy to get, the loan rates are usually much lower than other types of loans, and payback usually doesn't begin until the student's education has ended — and some students can even get their loans forgiven if they go into certain fields or agree to work in certain areas of need for a few years after graduation.

But you need to be careful with these loans to ensure that you don't end up owing more than you can afford to pay back. The loan of choice is a federal direct subsidized loan for undergraduate students. It's a fixed-rate loan with a current rate of 3.4 percent. You might also get a loan from a bank. One major bank currently offers fixed-rate loans for "as low as" 6.17 percent.

The mechanics of a student loan are about the same as for other types of loans. You borrow the money, and then you have to pay it back. To get a snapshot of payoff period and interest paid, visit `http://cgi.money.cnn.com/tools/studentloan/student loan.html`. You'll see, for example, that a $60,000 loan at 6.17 percent with $600 monthly payments will take 11 years and 9 months to pay off. You'll pay $24,457 in interest, too.

Calculating How to Avoid or Get Out of a Credit Card Hole

Decades ago, credit purchases were a novel way of making consumer loans. A family might have an account at the grocer's or general store. In the heyday of the great department stores, the store might issue a *charge-a-plate,* a small metal card that allowed on-account purchases at that store. Today, *bank cards* (notably Visa and MasterCard) are accepted practically everywhere.

Credit cards can be a great blessing or a great curse, depending on how you use them. In the following sections, I tell you how to avoid the trouble many credit card users get themselves into.

Understanding how credit cards work

A credit card is *revolving credit*. You have a credit limit, and you can borrow up to the limit. As you pay down your balance, you can borrow more. This sounds like a good deal — and it can be, as long as you pay off your balance each month. Otherwise, you must pay interest and may never pay off your balance. To see what happens when you don't pay off your balance, jump to the upcoming section "Paying down credit cards."

In addition, because the loan is unsecured, it usually has a relatively high interest rate, ranging from, say, 10 percent to as high as 33 percent. These high rates allow banks to make a lot of money, while offering you convenience in exchange.

When you use your card, the merchant pays a *swipe fee* for accepting it. You pay for this through higher prices.

Avoiding annual and other fees

Way back in the Stone Age, credit card lenders charged you an annual fee, typically $20.00, for the privilege of using their cards. Then annual fees went away for a long time. However, because of government activity (investigating and lowering swipe fees, for example), banks feel that they are hurting for money and may re-institute annual fees.

Other fees include inactivity fees (you have a card that you never use), application fees, cash advance fees, balance transfer fees, late fees, over-limit fees, return check fees, and foreign transaction fees.

A merchant may try to charge you a checkout fee (a surcharge) for using a credit card. Checkout fees are prohibited by law in ten states.

Your goal is to avoid, if possible, using any credit card that charges an annual fee. You can avoid most of the other fees, too.

Rewards, points, and folderol

Credit card lenders generally aren't too competitive over interest rates. Vast number of cards charge between 18 percent and 21 percent. To appear competitive, many banks offer what they call "rewards," "points," "loyalty points," or "travel points" as an incentive for you to use their cards. You can redeem those points for merchandise, travel, or in some cases cash. Here's a typical point setup: The Chase Amazon Visa, which I use, gives you 1 point (worth $0.01) to spend on Amazon when you charge $1.00 anywhere. The points accumulate all the time, and you can spend any number of them at any time.

There are two schools of thought about points. Some people swear by them, insisting that they give you something for nothing. Other people avoid them, believing that points muddy the water.

If you're in the "swear-by-them" crowd, make sure you use the points you accrue for things you actually need or will use.

Paying down credit cards

To enjoy the convenience of credit cards while avoiding the troubles, you have to use your credit cards wisely. Monthly interest accrues at 1/12 the annual percentage rate (APR). If, for example, the APR is 18 percent, then the monthly rate is 1.5 percent.

To figure your monthly balance, the bank takes your old balance, adds new charges, adds interest, subtracts payments, and shows you your new balance:

new balance = old balance + new charges + interest – payments

It's interesting to see how your balance grows when you make partial payments rather than paying the card off each month. Figure 10-8 shows seven months of credit card numbers with an 18 percent APR.

The lesson here is that if you charge $100.00 but make payments of just $50.00, eventually your ending balance will *grow*, not decline. If you have a big limit on your balance (for example, $16,000.00), the interest charges can get very dramatic as your balance approaches your limit (for example, $240.00 per month at the maximum balance for an 18 percent APR card).

Starting Balance	New Charges	Interest	Monthly Payment	Ending Balance
$0.00	$100.00	$1.50	–$50.00	$51.50
$51.50	$100.00	$2.27	–$50.00	$103.77
$103.77	$100.00	$3.06	–$50.00	$156.83
$156.83	$100.00	$3.85	–$50.00	$210.68
$210.68	$100.00	$4.66	–$50.00	$265.34
$265.34	$100.00	$5.48	–$50.00	$320.82
$320.82	$100.00	$6.31	–$50.00	$377.13

Illustration by Wiley, Composition Services Graphics

Figure 10-8: Looking at the increasing credit card balance.

Rule of thumb: Try never to charge anything that's consumed before it's paid for. That includes fancy dinners and movie tickets. After the thrill is gone, the payments linger on. The best thing you can do with a credit card — and admittedly it isn't easy — is to pay it off promptly.

Selecting Savings Accounts

Savings is a great tradition. The people in some countries save a relatively high percentage of their incomes. The United States hasn't usually been known as a nation of savers, but things are changing. Here are some of the savings options and the amount of interest each typically earns:

- **Regular savings account:** This is a "no strings" savings account. You "own" the money and can withdraw it all at any time. The disadvantage of a regular savings account is that interest rates are very, very low, as little as 0.01 percent.

  A *regular savings account* is sometimes called a *passbook savings account* because of the little book (the passbook) people used to carry to the bank. The teller wrote in the amounts of deposits and withdrawals in the book.

- **Certificates of deposit:** A *certificate of deposit* (CD) has a fixed term, such as 1 year, 2 years, 5 years, or 10 years. It pays better interest than passbook savings, and even better rates are available if you buy a jumbo CD, typically for an amount greater than $100,000.00.The drawback is that you're not expected to make any withdrawals. If you do, "severe penalties may apply." Your "math job" with certificates of deposit is simply to compare interest rates. The bank will do the rest for you. To compare rates, visit http://www.bankrate.com/. Typical rates are about 0.25 percent to 0.50 percent. A jumbo CD might yield 1.25 percent.

✔ **College savings accounts:** An *education savings plan* (a 529 plan) is designed to help families set aside money for college costs. It has tax advantages, because, although your deposits aren't tax-deferred, the money that grows in the account is free from federal (and sometimes state) taxation.

✔ **Medical savings accounts:** A *health savings account* (HSA) works "in partnership" with a high-deductible health insurance plan at your job. With a health savings account, you put aside money for medical expenses. The tax advantages include 1) not being taxed on money you put in, and 2) not being taxed on money you spend on qualified expenses. There are various limits and other strings attached.

Chapter 11

Key Principles of Investment Math

In This Chapter

▶ Determining your tolerance for risk

▶ Looking at retirement plans and pensions

▶ Investigating math related to stocks, bonds, and mutual funds

You are an investor. Maybe you're not as big as a giant pension fund or a Wall Street bank, but just the same, you're an investor. When you *invest,* you put money into something with the expectation of gain. The idea is that investments "grow money" until you need it. Food and clothing, while important, aren't investments. Savings accounts (covered in Chapter 10), stocks, bonds, and mutual funds are.

In this chapter, I make investment math as easy as possible for you. I explain the basic investment math, the vocabulary, and the tools (which include basic math skills, online calculators, and your fine mind) that can help you handle your everyday investment concerns.

Factoring Personal Info into Investment Decisions

No matter how much or how little money you have to invest, you need to know a little about yourself before you make any investment decisions, including how much to invest and what to invest in.

Knowing your tolerance for risk

Risk tolerance is your assessment of how much risk you're willing to take with investments. You've heard the adage, "There's no such thing as a free lunch." That's true of investments. An investment with a possibility of a higher payoff is likely to be riskier than one with a lower (but more reliable) return.

For example, if you put $100 in the bank at 1 percent interest and leave it there for a year, you know for certain that you will have $101 one year later. That's low-risk: You don't assume much risk, and your investment doesn't generate a big return. On the other hand, if you bet $100 on the flip of a coin, you have a 50 percent probability that you'll double your money ($200) and a 50 percent probability that you'll lose it all ($0.00). That's high-risk. Both the risk and the potential return are high.

Risk tolerance is mainly psychological. You need to determine whether you're risk-averse, risk-neutral, or risk-loving. If you deal with a broker, he or she will ask about your tolerance. Even without a broker, you can take risk tolerance surveys online. Visit http://www.paragonwealth.com/risk_tolerance.php.

 Conventional wisdom (and it's pretty good) links risk tolerance to your age. If you're younger, the thinking goes, you can afford more risk because you have time to make up for losses and you may see some good gains. If you're older, however, you typically need to be wary of risk to reduce the chance that your "sunset years" will become your nightmare years.

Looking at your investment horizon

Your *investment horizon* is the time in which you expect to buy and hold investments. This horizon is usually based on age. For example, if you're 60 years old and want to retire at 65, your investment horizon is about 5 years. If you're 25 years old, you've got a long time to save and invest. Your investment horizon is about 40 years.

Usually, those with longer investment horizons can concentrate on higher-risk investments, such as common stocks (or common stock mutual funds). Those with shorter investment horizons may be better off with fixed-income securities, such as bonds (or bond funds).

Choosing between appreciation or income

Based on your risk tolerance and your investment horizon, you are likely to pursue one of two broad categories of investments: those that grow in value (*appreciate*) and those that produce income.

✔ **Investments that appreciate:** *Appreciation* is an increase in the value of assets, including cash in savings and your home (you hope). When stocks appreciate, their price goes up, making them more valuable.

✔ **Investments that generate income:** *Income* from an investment, generally speaking, comes in the form of regular payments of interest or dividends. Income from investments is taxed as ordinary income, but that doesn't mean you spend it like ordinary income. Generally, you pay the taxes and plow the remainder into more investments.

✔ **A mix of investments:** At different points in your life, you may be advised to consider a mix of investments. When you're younger, you may have more aggressive growth investments and fewer income investments. As you grow older, you may shift to more conservative growth investments, bonds, and perhaps tax-free bonds.

Playing with Instruments: Not the Musical Kind

A financial *instrument* (sometimes called a *vehicle*) is an asset you can trade, and you'll be exposed to many of them over your lifetime. You can also call these assets *investment products*.

Basic financial instruments

Here's a brief tour of basic instruments:

✔ **Savings-based instruments:** These include savings accounts, certificates of deposit (CDs), treasury bills, treasury notes, and treasury bonds. They're considered savings-based instruments because you put money in and just let it grow. No buying or selling is involved. A savings account "goes on forever," and when a CD matures, you typically just renew it.

✔ **Investment-based instruments:** These include common stocks, preferred stocks, options, commodities, corporate bonds, and municipal bonds. These instruments require more involvement than savings-based instruments. The idea is to buy and sell them in a way that minimizes your losses and maximizes your gains. (By the way, a mutual fund is a collection of individual investment items. See the section "Managing Mutual Funds" for details on that investment vehicle.)

✔ **Real-estate-based instruments:** These include your home, a vacation home, a rental property, real estate investment trusts, or a whole apartment building (if you happen to own one). Even "raw" land is a real estate investment.

✔ **Other instruments:** This category includes just about anything — including gold, silver, and platinum — that doesn't fall into the other categories. Collectibles, such as art, stamps, coins, books, and a collection of Model T Fords, are instruments, too, in that they can increase in value and be traded for money, but the process of selling them can be slow. They're not very "liquid," as the saying goes.

✔ **Insurance-based instruments:** Some forms of insurance have an investment component. Many types of insurance offer only protection (think term life insurance or automobile insurance), but some insurances, such as whole life insurance and annuities, also have an investment component. I explain these in more detail in Chapter 12.

Climbing the investment pyramid

When you invest, pay a good deal of attention to conventional wisdom, which generally tends to be sound. Here's one example of conventional wisdom at work: Begin with the most reliable investments possible and, over time, add more speculative, risky investments.

Such a plan takes the shape of a pyramid. Figure 11-1 shows a typical investment pyramid.

If you follow a pyramid like the one in Figure 11-1, you can hold cash in some very reliable vehicles, such as a savings account and certificates of deposit. Even some insurance has conservative investment components that can return cash to you. After cash, your investments become riskier, but the payoffs are better. Mutual funds are pretty reliable, and so are bonds.

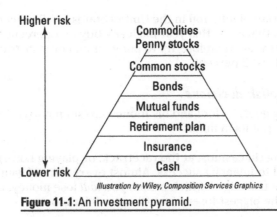

Illustration by Wiley, Composition Services Graphics

Figure 11-1: An investment pyramid.

A *pyramid scheme* is something entirely different from the investment pyramid. It's a fraudulent investment that you want to stay away from.

Doing a little at a time over time

How do you eat an elephant? One bite at a time. It's the same with investments. You don't need to save or invest a lot. Just put away *something*. Try $10 per week. A little money will grow a lot if you give it enough time. So the younger you are when you start, the better off you'll be when you're older. You can retire with a big pile of money if you work this right.

For example, if you put just $10 per week into a savings account paying 1.0 percent per year, at the end of the year (52 weeks later) you have $520 from your deposits and $3.04 in interest! (This example assumes monthly compounding, although banks may compound more frequently, but it shows how interest impacts your balance in a good way.)

It's crazy to think that you don't have to save a thing until you're about 60, and then you can build up your assets for retirement at age 65. Start early, even if you can only afford a little bit at a time.

Figuring in inflation

Conventional wisdom says that your investments have got to "beat inflation." That is, they have to earn at least as much as the rate of inflation. *Inflation* is the gradual increase in prices over time.

The current rate of inflation in the United States is 1.7 percent, which means that $1.00 this year basically buys 1.7 percent less than it did last year. Therefore, try to put your money in instruments that earn at least 2 percent.

Avoiding foolish decisions

As the saying goes, "A fool and his money are soon parted." Keep the following points in mind:

- Gambling (in casinos, at the racetrack, or playing Lotto) is *not* the road to financial success. Almost every form of gambling has a "house edge," which means you *will* lose money. Lotto is one of the biggest losers (see Chapter 9 for more).

- "Get rich quick" schemes won't make you rich. At best, they'll make someone else rich, with your money.

- If something is "too good to be true," it's probably not true. Beware of pitches that claim overnight doubling of your money, such as email solicitations from companies pushing *penny stocks* (those with a selling price of $0.01 or $0.02 per share).

Growing the Green Stuff: The Time Value of Money

Principal is the money you invest. *Interest* is the money that the principal earns. A savings account deposit, for example, is probably the simplest investment there is, and the bank pays you interest on your savings because it lends your money to others at a higher interest rate.

The *time value of money* is the value money, earning a certain amount of interest, grows to over a particular period of time. Over time, money that earns interest grows substantially. To see the way the time value of money works, take a look at simple interest compared to compound interest.

Calculating simple interest

Simple interest is calculated only on the principal invested. Say that your Aunt Tillie gave you $100 on your tenth birthday. Because your mother wouldn't let you spend it on frivolous things and told you to save it for a rainy day, you put it in a bank that pays 5 percent interest annually.

Here's how you calculate interest on $100 principal ($p$) at an interest rate ($i$) of 5 percent per year.

$$\text{interest} = \text{principal} \times \text{rate}$$
$$i = p(r)$$
$$i = \$100.00(0.05)$$
$$i = \$5.00$$

So, after 1 year, your $100 has earned $5. Every year, simple interest is the same. At the end of 50 years, your $100 will have earned $250 ($5 × 50) in interest.

By contrast, it's the world of compound interest where things get hot.

Calculating compound interest

Compound interest is calculated on the principal *and* interest accrued. Basically, you add the amount you made in interest to the principal. Then you let another period (usually 1 year) go by. You earn interest on the total. Then you do it again and again and again. With compound interest, from the end of year 1 to the day you withdraw, the total value of the investment climbs.

Given the same scenario as that used in the preceding section — your generous Aunt Tilly gives you $100 that your mother insists you save — you decide to put it in a bank that pays 5 percent interest *compounded annually.* Now say that 50 years have passed. How much money do you have in that old account now?

Figure 11-2 shows an example of the time value of money. Notice that at the end of year 15, your $100 has more than doubled to $207.89. At the end of year 50, your $100 is now $1,146.74 — a 1,000 percent increase! *That's* the power of interest.

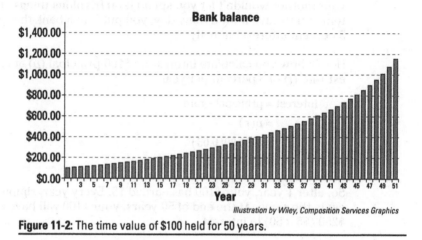

Bank balance

Year

Illustration by Wiley, Composition Services Graphics

Figure 11-2: The time value of $100 held for 50 years.

Albert Einstein said, "Compound interest is the eighth wonder of the world. He who understands it, earns it. . . he who doesn't. . . pays it."

Calculating future value

Future value is what your investment will be worth some time in the future; *present value* is what it's worth now. Using the Aunt Tillie example, in which the starting amount is $100, the interest rate is 5 percent, and the number of periods is 50 years, you can calculate future value with this formula. Here *FV* is the final amount, *PV* is the starting amount, *i* is the interest rate, and *n* is the number of periods:

$$\text{future value} = \text{present value} (1.00 + \text{rate})^{\text{number of periods}}$$

$$FV = PV(1+i)^n$$

$$FV = \$100.00(1+0.05)^{50}$$

$$FV = \$100.00(1.05)^{50}$$

$$FV = \$100.00(11.4674)$$

$$FV = \$1,146.74$$

The $100 will grow to $1,146.74.

Oh, see how it grows! The joys of making regular deposits

In the preceding section, I explained how compounding works and why it makes your money grow much faster. Now imagine that instead of a one-time deposit (the $100 in the earlier examples) you make regular deposits over a long period of time, as you would with monthly deposits you make to a college savings plan or regular deposits you make to any savings plan where you want to have a fixed amount by a certain date in the future (like a Christmas Club account).

To help you see how making regular deposits really boosts your bottom line, assume you have $100 per month to invest, and you're trying to decide whether to squirrel it away under your mattress or invest it. Let the info in the following sections help you decide.

Investment plan #1 — Money under the mattress

In this scenario, every month, you tuck another $100 under your mattress. To determine how much you'll have at the end of 50 years, use the following equation, in which *FV* is future value and *p* is the monthly payment:

$$\text{future value} = \text{monthly payment} \times 12 \text{ payments per year} \times 50 \text{ years}$$
$$FV = p \times 12 \times 50$$
$$FV = p \times 600$$
$$FV = \$100.00 \times 600$$
$$FV = \$60{,}000.00$$

If you hide $100 per month under the mattress for 50 years, that's 600 payments. The (now lumpy) mattress has $60,000 under it. Not bad!

Investment plan #2 — Money invested at 5 percent

In this plan, you put your $100 per month into an investment that pays 5 percent per year. To see how much you'll end up with in this scenario, use this formula, in which *FV* is the future value, *p* is the monthly payment, *n* is the number of payments (600 over 50 years), and *i* is the monthly interest rate (the annual rate of 5 percent divided by 12), which happens to be 0.004167.

$$FV = p\left[\frac{(1+i)^n - 1}{i}\right]$$

$$FV = \$100.00\left[\frac{(1+0.004167)^{600} - 1}{0.004167}\right]$$

$$FV = \$100.00\left[\frac{12.1194 - 1}{0.004167}\right]$$

$$FV = \$268,013.96$$

Plan #2 produces \$268,013.96, which makes it a much better deal than Plan #1. *Note:* The Excel spreadsheet formula for this calculation is =100*((1+0.05/12)^600–1)/(0.05/12).

These are pretty hairy formulas, right? To do more math with less effort, consider going to an online calculator, such as http://www.uic.edu/classes/actg/actg500/pfvatutor.htm. The key is to always use the best tool. When the math gets tough, check the Internet for a calculator.

Rounding Up Retirement Plans

A *pension* is a fixed sum of money you get, usually monthly, most often (but not always) on retirement from a job. An exception is that you may get a spouse's benefit, a widow's benefit, or children's benefit from the U. S. Social Security system. Here's the lightning round, summarizing retirement plans:

- ✓ **Employer plans:** These plans, known as *defined benefit* or *defined contribution* plans, are set up by employers for their employees, who have to meet certain eligibility requirements (like years of service and age) in order to receive the pension disbursements. A defined benefit plan tells you explicitly how much money you can look forward to in retirement. By contrast, a defined contribution plan has results based on how much money you put in. Small business employer plans are the SEP IRA and SIMPLE IRA.

- ✓ **Personal plans:** These are plans you create and maintain by yourself. Typically, the plans are the Individual Retirement Account (IRA) and Roth IRA. The benefit is based on how much and how frequently you put money into the plan. There may be certain tax advantages. To find out more, see the section "Calculating current and future tax advantages."

✔ **Government plans, such as Social Security:** Social Security is a defined benefit plan, financed through payroll contributions that you and your employer make. You pay a Social Security tax throughout your employment years, and when you reach a specified age, you receive payments from the Social Security Administration.

Some plans offer tax advantages. For example, in many plans, you don't pay taxes on a portion of your contribution when it goes in. Also, some plans have an "employer match," where the company puts in some money to match your contributions.

The feisty 401(k)/403(b): Defined contribution plans

Whether you work in the private sector or the public sector, your employer probably offers a defined contribution plan. A *defined contribution plan* is a retirement plan in which you decide how much money to put into it. What the account earns over time depends on how well the investment vehicle (one you choose from options offered by the company) does. Defined contribution plans have two key features:

✔ **Tax deferred:** Contributions are deducted from your paycheck before taxes are calculated. You'll pay taxes later on these contributions (probably at a lower rate) when you retire at about age 65.

✔ **Employer match:** A generous employer may offer to match part of your contributions. For example, some employers match 50 percent of the first 6 percent of your contribution. What this means is that, if you put in 6 percent of your gross pay, your employer will match 50 percent of that.

To figure the maximum employer match (based on 6 percent, for example), start by multiplying your annual salary by 0.06. Say you make $40,000.00 a year:

$40,000.00 \times 0.06 = $2,400.00

Then calculate 50 percent (0.50) of that amount.

$2,400.00 \times 0.50 = $1,200.00

You can still contribute more (often up to 10 percent of your annual salary). However, the employer stops matching at 50 percent of 6 percent.

Financial advisors point out that the employer match is "free money," and that you should take it.

A 401(k) plan is the type of retirement plan you find at corporations and medium-sized businesses. A 403(b) plan is the type of retirement plan you find at public education organizations and some non-profit employers.

If you "draw down" money early (which is allowed), you're looking at severe penalties, and you'll be paying income taxes at your current rate.

Adding up IRAs and their kin

Some retirement plans are very similar to 401(k)/403(b) plans. Almost all of them offer a tax advantage. IRAs and Roth IRAs are personal retirement accounts. SEP IRAs and SIMPLE IRAs are popular in the world of small business. Here's a summary:

- **Individual Retirement Account (IRA):** An IRA is your own personal retirement account. You currently can "tax defer" up to $5,000 per year (and $6,000 if you're 50 or older), meaning you pay no tax on that money now. When you make withdrawals at around age 65, you pay taxes on principal and interest at a lower marginal tax rate. See the section "Calculating current and future tax advantages." You can start making withdrawals between ages 59½ and 70½.

  Note: If you're covered by a retirement plan at work, your personal IRA tax-deferred deduction may be limited or reduced to $0.00.

- **Roth IRA:** A *Roth IRA* is a variation of the IRA. You pay taxes on your money *before* you put it into the plan, but you pay no taxes on principal or interest when you draw the money out.

- **Simplified Employee Pension Individual Retirement Arrangement (SEP IRA):** In a SEP IRA, you and your employer make contributions into a traditional IRA established in your name. There are tax advantages.

- **Savings Incentive Match Plan for Employees Individual Retirement Account (SIMPLE IRA):** Like the SEP IRA, the employee and employer make contributions to the employee's account, and there are tax advantages.

The dinosaur in the room: Traditional defined benefit plans

The defined benefit retirement plan is a vanishing entity. In a *defined benefit plan,* the company promises to pay you a specific amount every month after you retire. It's usually determined by your wages, length of service, and age.

These types of plans used to be the only retirement plans around. Unfortunately, many employers consider them to too costly and have replaced them with defined contribution plans. A few companies still have such plans, though. Municipal governments have them. Social Security and military retirement are defined benefit plans.

The biggest difference between a SEP IRA and a SIMPLE IRA is contribution limits. If you're an employee, your employer will have chosen one or the other. If you are a business owner, you'll need to get professional advice from a tax expert. You can also use Publication 560 from the Internal Revenue Service: `http://www.irs.gov/pub/irs-pdf/p560.pdf`.

Making sensing of Social Security

The United States Social Security program is the largest government program in the world. It's a retirement plan, with additional benefits for spouses, widows and widowers, children, and the disabled.

A key component of Social Security is that the plan is a defined benefit plan. Basically, you put your money into Social Security during your working years (via payroll deductions), and when you retire, Social Security sends you monthly checks.

What will your benefit be at retirement? Strictly speaking, the United States Social Security Administration (SSA) can't tell you that until you apply for benefits. However, the SSA provides a calculator to help you estimate your benefit in today's dollars or future inflated dollars. Visit `http://www.ssa.gov/retire2/AnypiaApplet.html`.

If you use the online calculator, be prepared to enter many years' worth of earnings.

Calculating current and future tax advantages

When it comes to taxes and retirement accounts, the U.S. government says, "You can pay me now, or you can pay me later." If possible, try to pay later. When you retire, you are likely to be subject to a lower *marginal tax rate* (also known as a *tax bracket*).

Those brackets are currently 10, 15, 25, 28, 33, and 35 percent. During their working years, many people earn enough to be taxed at the 25 percent rate. When you retire and start drawing down a retirement plan, you will likely be in the 10 percent or maybe the 15 percent bracket.

So what should you do? Defer your taxes now and pay them later, after you retire, or pay them now and get your retirement money free and clear? Well, free and clear always sounds good, but is it the smartest choice financially?

Say you have $50,000 in income and have to decide whether you want to defer taxes on $5,000 (a 10 percent IRA contribution) or pay them now.

If you pay taxes on the $5,000 now when you're in a higher tax bracket — say 25 percent — you'll pay $1,250 ($5,000 × 0.25 = $1,250). But if you defer those taxes until your retirement, when you will likely be in the 10 percent tax bracket, you'll pay only $500 ($5,000 × 0.10 = $500).

Deferring the taxes on $5,000 saves you money. Paying less in taxes seems like a better deal than paying more.

 Social Security has an earnings limit. If you continue to make money when you're receiving Social Security, your benefit will be reduced. However, the Social Security Administration says that withdrawals from your IRA don't count towards the Social Security earnings limit.

Managing Mutual Funds

A *mutual fund* is a collective investment. A mutual fund is a collection of, say, 30 or more securities. When you buy shares in a mutual fund, you own a little bit of all the securities within that fund.

At 2011's end, 7,581 mutual funds existed in the United States. With over 7,500 funds to choose from, be assured that they come in all flavors. Here are several major types of funds:

- **Stock (equity) funds:** Stock funds are made up of collections of common stocks. These are sometimes further broken into groups based on market capitalization (micro cap, small cap, mid cap, and large cap). That's a sign of the size of the companies.

- **Bond funds:** Bond funds are grouping of bonds. They include municipal bond funds and corporate bond funds.

- **Growth funds** and **income funds:** The names of these funds alert you to their purpose. Growth funds are collections of stocks with a likelihood of appreciating (increasing in price). Income funds are collections of stocks that have a good history of providing dividend income. For example, you may see funds with names like "Aardvark Small Cap Growth Fund" or "Aardvark Equity Income Fund."

- **Index funds:** Index funds try to imitate the performance and yield of well-known indexes, such as the Standard & Poor's 500 (S&P 500) index. The theory is that if the fund "matches the Dow" or "matches the S&P," then it's doing as well as the stock market in general is doing.

Paying attention to fees

Running a mutual fund costs money, and the investors "pay the freight." A fund's expenses may be met by including sales charges ("sales loads"), 12b-1 fees (annual fees), management fees, transaction fees, and anything else the fund managers can think of.

To complicate things, mutual funds usually have *classes* of shares: Class A, Class B, and Class C. These classes offer you choices in front-end loads, back-end loads, and 12b-1 fees:

- Class A shares usually charge a *front-end sales load* together with a small 12b-1 fee. I personally have paid 4.75 percent as a front-end load.

- Class B shares don't have a front-end sales load. They have a "contingent deferred sales charge" that declines gradually over several years. They are *back-loaded.* They have a high 12b-1 fee. Class B shares usually automatically convert to Class A shares after you've held them for a certain period, maybe six or seven years.

✓ Class C shares have no front-end or back-end load. They are *level-loaded* shares. How can this be? The load is in the high 12b-1 fee that you pay every year.

Some mutual funds are *no load funds,* which means there are no front-end or back-end sales loads. There may be various fees, such as annual "maintenance" fees. However, if you can find a true no-load fund with no 12b-1 fees, more of your money goes into the actual investment.

No-load funds are not automatically better. For example, a fund with a 5 percent front load that returns 15 percent annually outperforms a no-load fund with a 9 percent annual return.

To calculate a load, just multiply the amount of your intended purchase by the advertised load. If you want to buy $10,000 worth of a fund with a 5 percent load, calculate:

$$\$10,000.00 \times 0.05 = \$500.00$$

Some financial advisors consider Class C shares to be a marketing gimmick, trying to make them appear the same as no-load mutual funds.

Figuring the average annual return

Mutual funds are required to show their average annual return, which is an indicator of the fund's performance. The U.S. Securities and Exchange Commission says so. The fund has to show the Ending Redeemable Value (ERV) — the end value of a hypothetical $1,000 payment (the *model payment*) at the end of 1-year, 5-year, and 10-year periods.

Funds use the following formula to calculate ERV, where p is the model payment, t is the annual average return, and n is number of years:

$$ERV = \text{model payment}\left(1.00 + \text{average annual return}\right)^{\text{number of years}}$$

$$ERV = p(1+t)^n$$

$$ERV = \$1,000.00(1+0.05)^{10}$$

$$ERV = \$1,000.00(1.05)^{10}$$

$$ERV = \$1,000.00(1.62889)$$

$$ERV = \$1,628.89$$

In this example, the fund has an average annual rate of return of 5 percent. Over 10 years, the value of an initial model payment of $1,000 increases to about $1,630.

Considering Common Stocks

A *common stock* is also called an *equity*. Typically, a corporation issues shares that people can buy on the stock exchange, and the stock is said to be *publicly traded*. Other stocks besides common stock exist, such as preferred stock, but common stock is the basic form of equity.

Some people like to buy common stocks one security at a time. For example, they buy 100 shares of Intel or Procter & Gamble. Others like to buy shares in a stock mutual fund. See the earlier section "Managing Mutual Funds."

In any case, you usually buy stocks through your broker, who is well-credentialed and functions as a trusted advisor. A broker not only has credentials, but his company also has a research department, which is supposed to do good analyses; that's part of the reason the brokerage house charges commissions.

A *commission*, of course, is a fee you pay to a broker to execute a trade. It's often based on the number of shares or their dollar value. You pay a commission when you buy, and you pay another commission when you sell.

You can also buy stocks online. When you do so, you're on your own and must trust your own research and judgment. Trading online usually saves you money in commissions.

Processing P/E ratios

A stock's *price-to-earnings ratio* (commonly called the *P/E ratio*) is the relationship between a stock's share price and its annual earnings per share. By knowing the P/E ratio, you can determine if a stock is a good performer. A P/E ratio that's lower (a stock with a relatively low price and relatively high earnings) essentially means that you get a lot of earnings for the price.

For example, a stock that sells for $38.62 per share with earnings of $2.86 per share has a P/E ratio of 13.5. That's pretty good.

A P/E ratio between 10 and 17 is considered to be good. A P/E ratio greater than 25 suggests that a stock is possibly overpriced. A stock that sells at "high multiples" has attracted a lot of buyer interest (possibly irrational) and is sometimes called a *glamour stock*. For example, in 2011, a major fast-food chain had a P/E ratio of 52!

The easiest way to calculate a P/E ratio is not to calculate it. Let someone else calculate and display it. It's smart to visit websites such as Bloomberg: http://www.bloomberg.com.

Figuring your dividends

A *dividend* is a payment a corporation makes to stockholders. Usually, the company pays dividends four times a year. How much you receive in dividends depends on how well the company did and how many shares you own. If, for example, you own 100 shares and the dividend is $0.21 per share, you're going to get a check for $21.

If a company's stock doesn't pay dividends, that's okay. You (and the company's board of directors) may be more interested in the stock price going up (appreciating) than in dividends (income).

Dividend yield is the percentage relationship between your annual dividend and the price at which you bought a stock. High yields are considered better. For example, if a stock you bought at $10 per share pays $0.20 per share as a dividend, that's a lot better than getting the same dividend from a stock that cost you $50 per share. You get more bang for your investment buck.

To calculate dividend yield, you use this equation, where *i* is the dividends paid in a year and *p* is the market price:

$$\text{dividend yield} = \frac{\text{dividends paid in a year}}{\text{market price}} \times 100$$

$$Y = \frac{i}{p} \times 100$$

$$Y = \frac{(\$0.21 + \$0.21 + \$0.21 + \$0.21)}{\$34.99} \times 100$$

$$Y = \frac{\$0.84}{\$34.99} \times 100$$

$$Y = 0.024 \times 100$$

$$Y = 2.40$$

The dividend yield is 2.40 percent.

Investing in Bonds

A *bond* is a debt instrument. When you buy a bond, whether *corporate* or *municipal,* you're lending a corporation or a government entity a little money, and you'll receive interest until you sell the bond or it matures.

Bonds generate income, not appreciation. Municipal bonds are usually free from federal income tax. And if the bond is for an entity in your own state, it's probably free from state income tax, too.

Calculating interest on bonds

When you look up bonds on the Internet, you will likely see them described in a way similar to this:

CALIFORNIA ST PUB WKS BRD REGENTS UNIV CALIF-SER E, 5.00%, 04/01/2029

Translation? This bond is a California municipal bond (a *"muni"*) for the University of California system. It pays an annual interest rate of 5.00 percent and (if you don't sell it) will pay you a little money every quarter until April of 2029. Then you get all of your original investment back.

In other words, the money you make with a bond comes in the form of interest (tax-free interest, to boot, in this case). To calculate what one year's tax-free interest is, use this formula, where *i* is interest, *p* is principal, and *r* is the annual interest rate:

$$\text{interest} = \text{principal} \times \text{rate}$$
$$i = p(r)$$
$$i = \$1,000.00(0.05)$$
$$i = \$50.00$$

The example shows that a $1,000 bond earns $50 in interest in a year.

Calculating yield

Bond yield may be a little higher or lower than the interest rate. *Yield* takes into account whether the bond is selling at a little higher than $100 (people want it a lot) or a little less than $100 (people don't want it very much). If you pay a little more, you actually earn a little less. You'll still get interest at the "advertised" rate, but yield is thought to be a more accurate reflection of a bond's real return on your investment.

To calculate bond yield, you use this equation:

$$\text{bond yield} = \frac{\text{interest paid in a year}}{\text{market price}} \times 100$$

$$Y = \frac{i}{p} \times 100$$

$$Y = \frac{\$5.00}{\$107.778} \times 100$$

$$Y = 0.046391 \times 100$$

$$Y = 4.60$$

A single bond purchase usually requires that you spend a minimum of $5,000. Most people can't or won't do that. Instead, they buy into a corporate or municipal bond mutual fund, where the purchase amount is more flexible (sometimes as little as $1,000).

You use the same math to determine bond yield as you do for dividend yield.

Chapter 12

Covering Your Assets: Insurance Math

*L*ife is uncertain, and people usually don't welcome uncertainty. That's where the concept of insurance comes in. Insurance won't prevent bad things (an automobile accident, a house fire, or loss of life, for example) from happening, but it can reduce the financial loss you face should they happen.

The idea behind insurance is that you trade a small known cost (the payment, called an *insurance premium*) against a larger unknown cost. In this chapter, I tell you how insurance works and explain the math you need to know to understand what you're paying for. With insurance, the math consists of simple arithmetic you do and statistics that you don't do but should understand.

Honing in on How Insurance Works

Simply put, insurance insures you against a "peril," such as sickness, fire, auto accidents, or death. You make payments (called *premiums*) so that, if the peril happens to you, you can make a claim and the insurance company pays in the amount and manner specified in your policy. In some cases, the insurance company pays you money for some of your financial loss; in other cases, the company pays a service provider (the body shop you took your car to, for example, or the doctors and hospital you saw when you broke your leg).

When you're trying to decide how much coverage to buy and what you can afford, you first need to understand how insurance works — a task that takes you into the labyrinthine world of actuaries and risk assessment, which I explain in the following sections.

Here are a few terms you'll encounter in the world of insurance. Knowing them makes understanding the details easier:

- **Premium:** A *premium* is the amount of money charged for insurance. The term is used for all types in insurance.

- **Deductible:** A *deductible* is the amount of money that you pay from your own pocket to a service provider (doctor, mechanic, contractor, and so on) before an insurer pays a benefit. Deductibles are common in automobile insurance and health insurance. For auto insurance, you pay a deductible when you put in a claim for an accident. For example, if you have a $250 deductible on your auto insurance and you get into a wreck that causes $3,000 worth of damage, you pay the first $250 and the insurance pays the rest. With health insurance, over the course of a year, you pay your medical expenses yourself until you "meet the deductible." At that point, the insurer starts paying benefits. The next year, the deductible resets, and you have to do it all over again.

  Why do deductibles exist? If they didn't exist, the premiums would be out of sight. An insurer feels that, without deductibles, people would put in a lot of trivial claims.

- **Copay:** In health insurance policies, the copay is the amount of a healthcare bill that you pay when a service is provided. This isn't the same as the premium, which you pay to the insurance company to buy the insurance.

- **Co-insurance:** Co-insurance isn't exactly the same as a copay. Whereas a copay is a fixed amount (you have a $20 copay when you go to the doctor's office, for example), co-insurance is the percentage of the medical bill you're later responsible for. If the bill is $200 and your co-insurance amount is 30 percent, you pay $60 of the bill and the insurance company pays 70 percent ($140).

  Copays and co-insurance exist to prevent *moral hazard*. The purpose is to prevent people from seeking medical care that may not be necessary. The underlying philosophy is that, with no copay or co-insurance, you'll consume more care than you would if you weren't paying anything. Insurers believe copays are necessary to keep insurance costs down.

✔ **Lifetime maximum limits:** Health insurance traditionally had a lifetime maximum. A *lifetime maximum* was an expenditure cap. When you hit that amount, the insurance company wouldn't pay another dime. It's true that very sick people can run up big medical bills, but ultimately this limitation was perceived as too damaging to cancer patients and children with severely debilitating conditions. In the United States the Affordable Care Act has changed lifetime limits. So this is one term you no longer have to remember.

Spreading risk around

Insurance companies pay some or all of the costs people incur when a calamity strikes. Health insurance started this way — the insurance companies would protect patients from expenses associated with emergencies, injuries, or major illnesses, but other kinds of healthcare, like routine checkups or ordinary procedures, were simply paid for by the patients themselves. Health insurance has since absorbed the preventive and routine care portion and now contributes toward the payment of nonemergency healthcare expenses.

So here's the question: How can a business make money if it's responsible for paying for things that, by their very nature, tend to be very expensive? The answer is by distributing risk over a large pool of people.

For example, in a pool of 1,000,000 drivers, there may be 1,000 traffic accidents, making the probability that a driver will have an accident 1 in 1,000. The risk is spread over the whole pool. All 1,000,000 drivers pay a premium, but most drivers won't have an accident. Instead, the premiums paid by the accident-free drivers help pay the benefit for the drivers who do have an accident. Not a bad deal.

Evaluating risks

You don't know the exact chances of a peril happening to you, and your insurance company doesn't know either. But the insurer does know the statistical chances of it happening.

An *actuary* is a professional who analyzes risk and its financial impact. Many actuaries work at life insurance and health insurance companies, but they also figure risk for auto insurers. They can even determine when an appliance is likely to break down (and therefore what the manufacturer should charge for an extended warranty).

Here's why the actuary's job is important: The probability findings directly impact how much you pay for your coverage. Life insurance, for example, has always relied on life tables (also called *mortality tables*) developed by actuaries. It's 100 percent certain that a person will die, but no one knows when. A life table shows the *probability* that a person of a certain age will die before his or her next birthday. From that, an insurer can figure the remaining life expectancy for people at different ages and base premiums on life expectancy.

You won't be surprised to learn that after infancy, children have an excellent chance of living a long time. By contrast, people who are older (90 years of age, for example) have a poor chance of living another year. That's why healthy, non-smoking young adults can get life insurance for a song, and the elderly have to throw in a warm-up act, backup singers, and free backstage passes.

Probabilistic risk assessment (PRA) is a term for evaluating risks, such as those associated with an airliner or a nuclear power plant. PRA deals with the likelihood of a detrimental outcome of an activity versus the severity of the detrimental outcome. The *severity* (also called the *consequences*) is expressed numerically (for example, number of people hurt or killed, acres destroyed, dollars lost, and so forth).Think of PRA as asking "What's the worst that could happen and what are the chances?"

The *risk level* is basically the severity multiplied by the probability. The formula is

risk level = hazard severity × likelihood of occurrence

Life tables

For most of history, evaluating risk wasn't an exact science. Farmers guessed about the likelihood of crop failure, and kings guessed about the likelihood of war. But around 1662, John Graunt in London got scientific about death. He noticed predicable patterns when people in a group died. Graunt produced the first *life table*, based on a statistical summary of real experience. To this day, insurance companies still rely on life tables, updated of course.

Broadly, risk has these levels:

✔ **Low likelihood/low severity:** You can pretty much ignore these risks. There's not much chance they'll happen, and if they do, there's not much cost. Falling down the stairs would fall into this category. It's not likely to happen, but if it does, fixing the damage may require only first aid.

✔ **Low likelihood/high severity:** Risks that fall into this level aren't likely to happen, but when they do, you are in deep trouble. For example, an auto accident isn't a common occurrence, but it can produce severe bodily injury and costly property damage.

✔ **High likelihood/low severity:** These risks are moderate because, even though the chances of them happening are high, they usually don't have a hugely detrimental effect. You just have to cope with them. An example of this type of risk includes cutting yourself while cooking. Here, a good strategy is to try to reduce the risk (by being careful when you're slicing).

✔ **High likelihood/high severity:** These are top-priority risks and can mean big trouble. Those who live in areas subject to wildfires, hurricanes, and floods occasionally face this sort of risk and often have to evacuate their homes on very short notice. The best thing homeowners can do to minimize this risk is consider where they live. Flood insurance can help "hedge" against a flood and wind insurance may help with damage from hurricanes. For wildfires, you can reduce risk by keeping combustible brush at least 30 feet from the house.

Underwriters

In 1688, ship owners and merchants went to Lloyd's Coffee House in London to talk about their ships and the cargoes traveling on them.

A sea voyage is risky business — the ship owner or merchant can lose everything. A shipwreck can spoil your whole day. So bankers would accept some of the financial risk in exchange for a premium. They'd write their names under the risk information, and to this day those who accept risk are called *underwriters*.

Today, Lloyd's of London is a giant insurance company. And Lloyd's coffee house? There's a historical plaque on Lombard Street in London, and the original storefront is on display at the National Maritime Museum.

Determining premiums

An insurance *premium* is the amount you pay for coverage. Premiums vary a lot, depending on a number of factors:

- ✔ **The degree of "peril" you face:** For example, as you grow older, life insurance costs more because the risk of dying is higher. The same is true if you are in a risky profession. With other lines of insurance, premiums vary if the insurer believes you are at greater risk. For example, a health insurer may want to charge you an increased premium if you're a tobacco user. Teen drivers pay more for auto insurance than older, more experienced drivers.

- ✔ **The cost of claims:** Premiums grow because the cost of claims grows higher and higher. In the world of auto insurance, your premium may go up if you get into an accident. In health insurance, premiums rise regularly because of the increasing costs of healthcare. Also premiums for homeowners around the nation jumped up following Hurricane Katrina due to the costs the insurance companies incurred in the aftermath of that disaster.

- ✔ **How much coverage you buy:** A $20,000 life insurance policy costs twice as much as a $10,000 life insurance policy, for example. With auto insurance, the premium also depends on how much coverage you buy. For example, getting full coverage insurance costs more than getting only liability insurance.

In all cases, the insurance company works out the statistics. Your real-life math job is to examine the coverages and compare the premiums. I tell you how in the next sections.

Making Calculations about Your Own Lines of Insurance

Life insurance, health insurance, automobile insurance, and homeowner's insurance are the major personal lines that people buy.

With some insurance, you can pick the amount of coverage and then compare premiums. Life insurance is sold in units (typically units of $1,000, the total making up the *face amount* of the policy), auto insurance is sold in varying amounts, and health insurance has plans with different deductibles and different benefits. Your math task related to insurance is usually to compare the coverage offered to the premiums charged and pick the plan that works best for you.

A *personal line* of insurance covers individuals and their property. A *business line* of insurance covers business property (such as buildings and ships) and business activities (which may cause a professional liability). To split hairs, the health insurance your employer offers you is a business line, sold to the corporation by a corporate insurance agent.

Auto insurance

When you look for auto insurance (or evaluate the plan you already have), you need to consider things like the value of the vehicle you're insuring and how much you can afford to pay as the consequences of an accident. Both affect the kind and amount of insurance you need and the amount you'll end up spending for it.

Some coverages levels have minimums, mandated by the state you live in, but others coverages are flexible and will affect how much or how little you pay.

When you review your auto insurance policy or look for a new policy, you'll see separate charges for up to six items: property damage, bodily injury, collision, comprehensive, medical, and uninsured motorist coverage. There's also a limited form of coverage called PLPD (personal or public liability and property damage insurance).

Value of your vehicle

What's your car worth to you? Chances are it's worth a lot less to the insurance company. Collision insurance pays for damage to your car (damage to the other driver's car is covered by property damage insurance). Collision insurance also pays if you drive your car into a tree or your other car (as I did once).

The amount of collision coverage will be "actual cash value less deductible." After an accident, your insurer determines your car's value and what it would cost to repair it; you can only select the deductible. (Of course, collision coverage may be optional, but you're taking your chances if you opt out.)

Say your $22,000 car was worth $6,000 before an accident. If there's a lot of damage (a total loss), the car may not be worth fixing. If the salvage value of the car is $500, the insurer doesn't have to pay more than $5,500 in repairs. The company can take the car and just give you a check for $5,500. That's what is meant by your car being "totaled."

TIP

Certain cars — performance cars, luxury cars, and cars that are most often stolen, for example — cost more to insure. Before you buy a new car, call up an insurance agent or go online to see what the going rate for insuring that kind of car is.

Amount of deductible

Sure, having a low deductible sounds great. After all, wouldn't it be nice to not have to shell out anything (or shell out very little) in the event of an accident? However, low-deductible coverage costs more. So you need to balance the premium cost with the deductible amount before deciding.

Auto insurers will not give you a cost comparison of deductibles versus premiums online. To make a comparison, you usually need to talk over the phone with an agent. When you see the impact of different deductibles on premiums, then you can subtract to see the savings from a higher deductible.

Homeowner's insurance

Homeowner's insurance is useful for insuring your house against fire, theft, and a giant tree falling on the roof. Homeowner policies usually have some liability coverage ("slip-and-fall insurance") in case someone's injured on your property. Renter's insurance is a variation of homeowner's, intended for renters.

When you're buying insurance to cover theft or destruction of personal property, you may have a choice between *actual cash value* or *replacement cost*.

✔ **Actual cash value:** With actual cash value, in the event of a loss, the insurance company reimburses you the amount that the item is worth today. For example, if you have a lightning strike that wipes out your 3-year old TV, your 15-year old dishwasher, and your 1-year old computer, you get only what those items are worth today, regardless of what you paid for them new and how much they would cost to replace now. Bottom line: You probably couldn't replace the lost items for what the insurance company will pay.

✔ **Replacement cost:** With replacement cost, the insurance company reimburses you for the amount that the lost item would cost if you were to buy a similar item at today's prices. As you might imagine, replacement cost is the more expensive coverage.

When you're insuring personal property, you want to *value* the items you're insuring. Make a list showing what you own, when you bought it, and how much it originally cost. (Digital photos are helpful, too, in case you have a dispute with the insurance company.)

Health insurance

Health insurance is trickier to understand than other kinds of insurance because of how complicated it is, due to various types of coverage, various deductibles, discounts negotiated between the service provider and the health insurance company, copays, co-insurance amounts, and so on.

The best real-life math approach is to understand your health insurance policy (which isn't always easy) and to keep track of your medical expenses. The following sections tell you what kind of info you have to wade through to make sense of your medical bills and how you can anticipate your share of the totals.

Health insurance doesn't insure your health. It insures you against a portion of your financial losses. The only real "health insurance" comes from diet, exercise, and regular doctor visits when you're well. Be sure not to forget prenatal care, either.

Deciphering your bill

Even highly intelligent people can go insane trying to decipher their medical bills. Take a look at Figure 12-1, which shows a health insurance explanation of benefits.

PLEASE REFER TO THE BACK OF THIS DOCUMENT FOR ANY ADDITIONAL INFORMATION

Type of Service	Service Date	Code	Total Charges	PPO Discount	Not Covered	Covered Amount	Copay	Balance	Total Payment	See Remarks
OP DRS VISIT	06/26/2012	99213	107.00	58.25	48.75					RR PH AG 08
OP RADIO & LAB	06/26/2012	93000	31.00			31.00				RR 12 AG 08

IRS Withholding:
Adjustment Due to Other Insurance:

Total Billed Charges: $138.00

Total Paid:
Patient Liability: $79.75

(Patient Liability includes all applicable copayments, co-insurance, deductibles and non-covered items)

PAYMENT SUMMARY SECTION

Payment made to:
Amount Paid:
Check Number:

PPO Discount Total: $58.25
Base Deductible:
Optional Benefit Deductible: $31.00

Illustration by Wiley, Composition Services Graphics

Figure 12-1: Health insurance explanation of benefits statement.

What you may have noticed with health insurance explanations of benefits is that they aren't necessarily easy to understand. Here's what the fields in an explanation of benefits statement (EOB) mean (keep in mind that the EOB you receive may differ):

- ✔ **Total charges:** This reflects what the doctor charges for a service.

- ✔ **PPO discount:** This is amount that the doctor agreed to let go of because his or her practice is part of a preferred provider organization (PPO) network that negotiated a lower fee for that service. (This is one reason why seeing a doctor who's "in the network" is supposed to make your payments lower.)

- ✔ **Not covered:** Some services (or portions of a service) aren't covered by your policy. In this statement, $48.75 of the $107 visit wasn't covered by insurance. You're responsible for this amount.

- ✔ **Covered:** The insurance company will normally pay all $31 of the lab charges.

- ✔ **Optional benefit deduction:** Even though the $31 lab fee is covered, you haven't met the deductible, so you're paying the $31.

- ✔ **Patient liability:** You owe $79.75, and that's the fact. It's a combination of $48.75 for the visit and $31 for the lab work.

- ✔ **Code:** This is the procedure code; it's the way the doctor's office indicates what kind of service was provided. You'll never find out what these codes mean without the Internet — and it's not on the back of the explanation. In the example, *99213* stands for "15 minute visit with established patient."

Although every insurance company has its own explanation of benefits form, they all use terms similar to those in the preceding list. Understand those, and you're that much closer to understanding your medical bills. Your real-life task is to tear out your hair, subtract what you can, and add up what's left. You then wait for the doctor's bill (which will reflect the insurance company's payment) and pay the balance. For a specific rundown of what your EOB or billing statement shows, call your healthcare provider's billing office.

Choosing between high premiums or high deductibles

Plans with lower deductibles are more expensive. "Cadillac plans," for example, have generous benefits and lower deductibles; therefore, their premiums are higher. To keep premiums affordable, you may want to choose a plan with a higher deductible so that you can reduce the amount of money you pay in premiums to the insurance company.

However, a high deductible/low premium plan isn't always the best choice, depending on your circumstances. To determine whether a high deductible/low premium plan is better for you than a high premium/low deductible plan, you need to assess how likely you are to need medical care during a year and at what point the amount you save in deductible offsets the amount you pay in premiums.

Say, for example, that you have to choose between a plan with a $2,000 deductible that costs $600 per month and a plan with a $5,000 deductible that costs $400 per month. The difference in premiums is $200 per month, and the difference in deductibles is $3,000. Your job is to save $3,000 to make up for the difference in deductibles. So divide the $3,000 by the $200 monthly premium savings to see that after 15 months, you'll have enough in the bank to make up for any difference *for a major illness*. For minor things, such as doctor visits and lab procedures, don't try. You probably won't "meet the deductible" in either case.

Life insurance

Life insurance comes in many forms, mainly whole life (cash value) insurance, term life insurance, and annuities. When you're employed by a large company, chances are that it may offer you free term life insurance equal to your annual salary, with the option of buying two or three times that amount at a low rate.

Term life insurance insures your life for a fixed period of time, such as 5 years, 10 years, or 20 years. It's fairly low cost. To determine your premium, simply multiply the number of units you want (say one hundred $1,000 units of 5-year term life insurance) by the annual premium per unit (say $1.00). In this example, the annual premium is $100.

Sometimes, the insurance company simply advertises its total premium. For example, you may see "$250,000 coverage, 20 years, $177.50 annual," and that pretty much tells you the whole story.

Whole life insurance insures your life until you die, which (for five-year-olds) may be 90 years in the future. Even though your chances of dying increase each year (and the premium should increase as well), people prefer *level-premium insurance*. That is, the premiums stay the same over the entire life of the insurance contract. That means you pay a little more than necessary in the early years and a little less than necessary in the later years. Over time, an excess accumulates, and it's known as the *cash value* of the policy. Again, to determine your premium, simply multiply the number of units you want (say ten $1,000 units of whole life insurance) by the annual premium per unit (say $6.00). In this example, the annual premium is $60.

In the case of life insurance, you may buy additional coverage, called *riders*. These include spouse riders, mortgage payoff riders, and guaranteed insurability option riders.

How do I insure thee?
Let me count the ways

If you look at all the insurances available, your head will spin, a condition not usually covered in health insurance policies. Here's the lightning round of some forms of insurance other than those covered in this chapter.

✔ *Private mortgage insurance (PMI)* is also known as *lenders mortgage insurance.* It's insurance against your defaulting on your home loan. The bank gets the benefit and you pay the premium. You can read more about this in Chapter 10.

✔ *Unemployment insurance (UI)* is paid by your employer. You get a benefit should you lose your job through no fault of your own. Note that UI is insurance, not a handout or any form of "welfare."

✔ *State disability insurance* is insurance you pay through payroll deduction. It pays medical benefits for on-the-job injuries.

✔ *Non-medical health insurance* (if that's not too much of a contradiction) includes disability income insurance, long-term care insurance, and Medicare supplement coverage (Medigap).

✔ *Pet insurance* has grown in popularity. You can insure your pet against illness, accident, or death.

✔ *Cruise insurance* protects your investment in a cruise if, say, a hurricane or other travel disaster prevents you from going or prevents the cruise from taking place. For small trips, it's not so important, but for a $20,000 once-in-a-lifetime cruise, buying cruise insurance is probably an excellent idea.

Chapter 13

Taking Math to Work

· ·

In This Chapter

▶ Calculating markups, discounts, and time and material costs

▶ Getting familiar with the different kinds of profit

▶ Filling out your timesheet and project accounting sheet

▶ Comparing what you make to what you take home

· ·

*H*ow much math does your job require? For better or worse, most jobs don't require much math. The exception is when you're a bank teller, an accounting clerk, a retail clerk, or a chief financial officer — and even then a lot of the math is "hidden" from you. It's handled inside the cash register or by the proprietary accounting software your company uses.

Even if you don't do math on the job, knowing a bit about business math is very helpful. Business math encompasses everything from the high-flying items, such as complex calculations that should be left to experts, to the low-flying items that everyone who earns a paycheck should know. Guess which one I cover in this chapter? (*Hint:* Think of the topics as an incredibly brief course in Business Math Lite.)

Setting Prices

Profit comes from the difference between wholesale and retail prices. *Wholesale prices* are the prices that the business spends to buy items. Retail prices are the prices the business charges for items. For example, a business may buy a widget from a wholesaler for a price of $0.50 (the wholesale price) but turn around and sell the widget to customers for $2.00 (the *retail price*).

Stores offer *discounts* (the difference between an item's regular price and its sale price) in order to improve sales. Although the business charges less, it can make up that "lost" profit because people love sales. Discounting merchandise (the theory goes) will increase a store's volume of sales. While they are in the store, customers might even buy items that aren't on sale.

These two concepts — setting retail prices and offering discounts — give rise to special calculations that you need to know about, whether you're on the buying end of the transaction or the selling end.

Managing markups

A *markup* is the difference between the cost of goods and services and their selling price. You express the markup as a percentage of the wholesale price. Of course, the selling price is supposed to be high enough to create a profit.

To calculate a product's markup, you subtract the wholesale price from the selling price and then divide the result by the wholesale price. Suppose, for example, that the wholesale price for an item is $1.50 and the retail price is $2.50. You can use this formula to calculate the markup percentage:

$$\text{markup} = \frac{(\text{selling price} - \text{wholesale price})}{\text{wholesale price}} \times 100$$

$$\text{markup} = \frac{(\$2.50 - \$1.50)}{\$1.50} \times 100$$

$$\text{markup} = \frac{\$1.00}{\$1.50} \times 100$$

$$\text{markup} = 0.67 \times 100$$

$$\text{markup} = 67$$

An item that costs $1.50 and sells for $2.50 has been marked up 67 percent.

If the boss asks you to mark up a new item by 30 percent, use this formula: wholesale price × (1 + markup). In this example, the wholesale price is $1.50:

$$\text{selling price} = \text{wholesale price} \times (1 + \text{markup})$$

$$\text{selling price} = \$1.50 \times (1 + 0.30)$$

$$\text{selling price} = \$1.50 \times 1.30$$

$$\text{selling price} = \$1.95$$

When you mark an item with a wholesale price of $1.50 up by 30 percent, you sell it for $1.95.

Figuring discounts

Discounts lower the selling price of goods or services.

The formula for figuring a discount is way simple. Just multiply the regular price by 1 minus the discount rate. Suppose, for example, that you want to discount an item that regularly costs $2.50 by 20 percent. Use this formula:

$$\text{sale price} = \text{regular price} \times (1 - \text{discount rate})$$
$$\text{sale price} = \$2.50 \times (1 - 0.20)$$
$$\text{sale price} = \$2.50 \times 0.80$$
$$\text{sale price} = \$2.00$$

When you discount an item 20 percent, you sell it for 80 percent of its regular price. If you apply a 20 percent discount to an item with a regular retail price of $2.50, its sale price tag is $2.00.

Quantity pricing produces lower prices, too, but it's not a discount. See Chapter 5 for details.

Predicting time and materials costs

Time and materials (also known as *labor and materials*) is a standard phrase in some construction contracts. The customer agrees to pay labor rates for actual hours worked and for the actual costs of materials used. It's open-ended, the opposite of a fixed-price contract. Automobile mechanics and plumbers do something similar. Your bill reflects materials used and labor hours expended.

Some contracts guarantee a maximum price, which puts a limit on the total charges. At my company, we call it an NTE ("Not To Exceed") price.

The formula for calculating time and materials is very simple: You simply determine the price for the time (multiply the hours worked, or estimated to be worked, by the per-hour price) and then add in the price of parts and any applicable sales tax.

Bumping into business numbers

Various business numbers have names. If you know the names and the basic calculations to derive them, you'll be an expert in no time at all. (Or at least you'll appear to be an expert.)

✔ **Income:** Income (sometimes called *revenue*) is money your company takes in from selling goods and services. Selling an asset (like a truck) isn't income.

✔ **Expenses:** The term *expenses* refers to the money your company spends to make products and deliver services. Expenses are also called *costs*. Buying an asset (like a truck) isn't an expense. True, the purchase reduces the company's cash, but it gets a truck.

There are two kinds of costs: direct and indirect:

✔ **Direct costs:** These costs are tied directly to making a product or delivering a service. Direct costs are pretty easy to figure out. Each product your company makes (for example, a pair of denim pants) requires a certain dollar amount for materials (for example, denim, thread, zippers, and buttons) and labor. Direct costs make up the cost of goods sold.

✔ **Indirect costs (also known as *overhead*):** These costs aren't directly applicable to the product itself and include things like administration, personnel, and vehicles. Indirect costs make up operating expenses.

Consider this automobile repair example. Here, the hours are worked out to the hundredth of an hour, and the labor rate is figured at three decimal places:

$$\text{total charges} = (\text{hours} \times \text{hourly rate}) + \text{parts} + \text{tax}$$
$$\text{total charges} = (4.17 \times \$89.928) + \$390.40 + \$30.74$$
$$\text{total charges} = \$375.00 + \$390.40 + \$30.74$$
$$\text{total charges} = \$796.14$$

By the way, the example total is exactly what it costs to replace a timing belt on a 2005 Honda Civic.

Paying Attention to Profit

Businesses are in business to make money — a concept even kids selling lemonade in their front yard understand. But whereas little Molly is happy to end up with a profit of a few extra dollars at the end of her lemonade-selling season (and gives it no more thought than what she can spend it on the next time Mom takes her to the store), businesses have to parse their profit a little more finely.

Calculating profit margin

The *profit margin* (or just plain *margin*) is the net income divided by the selling price (also called *revenue*). The *net income* for an item is the selling price less the wholesale price. To calculate the profit margin, use this formula:

$$\text{margin} = \frac{(\text{selling price} - \text{wholesale price})}{\text{selling price}} \times 100$$

$$\text{margin} = \frac{(\$2.50 - \$1.50)}{\$2.50} \times 100$$

$$\text{margin} = \frac{\$1.00}{\$2.50} \times 100$$

$$\text{margin} = 0.40 \times 100$$

$$\text{margin} = 40$$

When you sell an item with a wholesale price of $1.50 for $2.50, the net income is $1.00. The profit margin is 40 percent.

Determining gross profit

Gross profit is sales less the cost of goods sold. In a service company, such as a landscaping service, the cost of goods sold includes the cost of labor to do the landscaping. Here's the formula:

$$\text{gross profit} = \text{sales} - \text{cost of goods sold}$$

If that was the end of it, everything would be fine, but there's more.

Pre-tax profit

Pre-tax profit (sometimes called *operating profit*) is what you get after you subtract all the operating expenses from the gross profit. Here's the formula:

pre-tax profit = gross profit – operating expenses

The operating expenses include overhead. See the sidebar "Bumping into business numbers."

Net profit

Net profit is the logical end to figuring the company's bottom line. To get net profit, you subtract taxes and interest from pre-tax profit.

net profit = pre-tax profit – taxes – interest

Making Change

No matter what your career, the realities of business are that you buy and sell. If you buy for cash, you receive change in return. If you sell for cash, you're expected to give change. Either way, knowing how to make change accurately is a good idea.

Making change is the technique of returning to a customer the difference in cash between the amount of a purchase and the money tendered. *Change* refers mainly to loose coins, but in reality it regularly includes paper money.

The modern school of change-making uses a "no-math" technique. Basically, the cash register tells the clerk the charges. You give the clerk money, and he or she enters the "amount tendered" into the register. The cash register tells the clerk what change to give you. This method is usually fast, efficient, and accurate, but if the register isn't working, you will sometimes see the clerk struggle. If you don't know how to count change, you'll struggle as well, and it's anybody's guess whether you're getting back what you're owed. In addition, some "businesses" — church or school bake sales, for example — won't have automated cash registers.

What do you do? You count out change the way your grandmother did when she worked at Woolworth's decades ago. Say your customer makes a $9.56 purchase and gives you a $10 bill.

1. **Leave the bill in plain sight on top of the cash drawer or cash box.**

2. **Count the change out and give it back by "Speaking the change."**

   In this method, you describe what you're giving back. Start with the smallest coins.

   "Your purchase was $9.56, out of $10.00. That's $9.56,"

   "plus 4 cents (4 pennies) makes $9.60,"

   "plus 5 cents (a nickel) makes $9.65,"

   "plus 10 cents (a dime) makes $9.75,"

   "plus 25 cents (a quarter) makes $10.00."

No confusion, no error, and — if you'll notice — no math!

Tracking Your Time

Your time is the most precious commodity you have. On the job, you'll see many measurements of time. Management measures the hours of operation of machine tools (for maintenance) and the time to produce items (for productivity). If you're a person on the job, you have two key elements of time to track: the hours you'll be paid for working and the hours you expend on different projects.

The timesheet

Most people are paid by the hour. They're called "non-exempt employees," because they aren't exempt from overtime laws (and that can be a very good thing).

A *timesheet* is a common form used to record the time you spend doing your work. These forms come in many variations (and if you've worked at several different companies, you've undoubtedly seen several "flavors"). Figure 13-1 shows a simple timesheet.

The key task is to record your time accurately and do a little bit of addition.

Willow Valley Software
562 Brock Road
Nevada City, California 95959
530/265-4705

PROFESSIONAL SERVICES TIMESHEET

NAME: Joe Writer   WEEK ENDING DATE: 04/12/2012

PAYROLL

	MON	TUE	WED	THU	FRI	SAT	SUN	TOTAL
Regular hours	8	8	8	8	8			40
Overtime hours								
Vacation hours								
Holiday hours								
TOTAL	8	8	8	8	8			

TOTAL PAYROLL HOURS   40

Illustration by Wiley, Composition Services Graphics

Figure 13-1: A timesheet.

To complete a simple timesheet, you generally have to record the hours you work each day in the appropriate box or field and then add them up to give the total hours each day, the total hours each week, and so on. Timesheet math is all about addition. For the timesheet shown in Figure 13-1, for example, follow these steps:

1. **Enter the number of hours you worked each day, separated into the given categories.**

2. **Add up each day's hours and place the total at the bottom of each column.**

3. **Add across each row to total each category and place that number in the right-hand column.**

4. **Enter the total for the week.**

   The sum of each day's hours should agree with the sum of the category hours.

Note that this timesheet has room for overtime hours, but the company doesn't generally authorize overtime. For more on overtime, see the section "Calculating your gross pay."

You may work in a place that has a time clock. You "clock in" when you arrive and "clock out" for lunch. Then you "clock in" after lunch and "clock out" when you leave. The way to figure out the morning and afternoon hours so that you know how many hours you worked in a day is to subtract times. If you clocked in at 8:00 a.m.

and left for lunch at 12:30, for example, you figure the difference: 4.5 hours. Then, to figure your afternoon hours, you subtract your ending time (say, 5:30 p.m.) by the time you clocked back in from lunch (say, 1:30): 4 hours. Add your morning and afternoon hours together to get your total hours worked (8.5 hours). (*Note:* Many time clocks use military time, but the math is the same.)

Heavens, no — Not project accounting!

Heavens, yes! You always report time on the job for payroll purposes, but often you're also required to do project accounting. *Project accounting* allows management to track which projects consume employee time. (This information is often necessary if projects are budgeted for a certain total number of hours to complete.) Figure 13-2 shows a project accounting entry sheet. Notice how similar it is to a timesheet. Basically, the task is the same; you just separate each day's hours differently. Rather than categorizing hours as regular, overtime, and so on, you categorize them by project (the cryptic codes in the left-hand column in the figure, for example). Project accounting math is easy. Just record correctly and add.

NAME: Joe Writer	PROJECT ACCOUNTING						Week Ending: 04/12/2012	
PROJECT	**MON**	**TUE**	**WED**	**THU**	**FRI**	**SAT**	**SUN**	**TOTAL**
CIWMB Training	8	8	4					20
CIWMB Editing Services			4	4	2			10
CIWMB Assistance				4	6			10
HP – CASL								
HP – XP								
TOTAL	8	8	8	8	8			

TOTAL PROJECT ACCOUNTING HOURS | 40

Illustration by Wiley, Composition Services Graphics

Figure 13-2: A project accounting sheet.

In this example, Joe Writer spent all of Monday, all of Tuesday, and half of Wednesday on CIWMB Training. The rest of his 40 hours worked were spread between two other projects.

The Total column at the right sums up hours spent on each project. The Total row at the bottom sums up each day's hours. The sum of each day's hours should agree with the sum of the project hours.

Accrual versus cash accounting

The two major accounting methods are the *cash method* and the *accrual method*:

Cash method

In cash basis accounting, business activity at any given moment is determined by when cash actually flows into or out of a business. Cash basis accounting is simple. Basically, the business checkbook tells the whole story. When you pay for materials for your business with a check, that's cash out the door. When you receive money for selling your product, that's cash in the door. Small businesses often use cash basis accounting.

The trouble is, cash basis accounting ignores timing. If you buy materials by credit card, you use them now, but don't pay until later. If you "bill out" clients, you're selling to them (or working for them) now, but they don't pay until later. These amounts don't show up immediately on the business ledger, even though they're vital to understanding business activity.

That's where accrual basis accounting comes in.

Accrual method

In *accrual basis accounting,* the financial picture of a business at any given moment is based on when income and expenses are actually incurred. Accrual basis "realizes" expenses the moment you incur them, not when you pay for them. It "realizes" income the moment you sell a product or service, not when the customer pays.

Here's how it works: When you create an invoice, the accounts receivable (A/R) system generates a receivable, even though the customer may not pay for, say, 30 days. When the payment comes in, the receivable "goes flat." It's been satisfied by the payment. Accrual basis accounting used to be hard, but now some great accounting programs make it easy.

Accrual basis accounting is thought to provide a more accurate reflection of business activity than cash basis accounting. Most businesses use accrual basis accounting.

Parsing Your Paycheck

You know that your salary isn't a gift. It's the money you get for expending your time and energy on work. Your paycheck returns some of that energy to you.

If you work 40 hours a week, for example, and are paid $20 per hour, your *gross pay* is $800. If you work overtime, you can earn even more gross pay. Of course, you don't bring home everything you earn. The taxman cometh, and his deductions, as well as other deductions your employer implements (like health or life insurance deductions), reduce your paycheck.

In the following sections, I explain how to figure your gross pay when you get paid regular time, overtime, and double time wages, and how to figure what you can expect to take home after deductions.

Calculating your gross pay

It's a fact that most people work hard to earn their salaries. If you're an hourly employee, here are the major classes of time expended on the job:

- ✔ **Regular time (also known as *straight time*):** Regular time is usually time paid on the first 40 hours you work in a week, at your base rate.

- ✔ **Overtime:** When you work more than 8 hours in a day or more than 40 hours in a week, hours beyond the 8 hours or 40 hours are paid at the overtime rate, which is often "time and a half." What this means is that you earn your base rate plus half your base rate for every hour of overtime hour you work. Say your base pay is $20 per hour and you work 10 hours of overtime. Your hourly overtime rate is $30 per hour.

- ✔ **Double time:** When you work more than 12 hours a day, hours beyond 12 hours are paid at the double time rate. (In Hollywood, this is called "Golden Time.") With a base pay of $20 per hour, you make $40 per hour when you work double time.

- ✔ **Sick time:** Sick time is a company benefit. The company pays you straight time for the hours you were out sick.

- ✔ **Vacation time:** Like sick time, vacation time is a company benefit. Many companies pay 80 hours per year in vacation pay.

Figuring your net pay: All about deductions

Net pay isn't the full amount of your paycheck. It's the amount you can spend. Whereas your *gross pay* (see the preceding section) tells you how much you've earned based on the hours you've worked or the salary you negotiated (if you're an exempt employee), net pay is essentially what's left over after deductions.

Payroll deductions come in two flavors — *mandated* and *voluntary*. Your employer *must* withhold some money for federal income tax, Social Security and Medicare taxes, and other mandatory items, like state income taxes and state disability insurance, where applicable. You decide on voluntary deductions. They may include retirement plans and charitable contributions.

Figure 13-3 shows typical payroll deductions. You probably recognize most of them:

- ✔ **Federal income tax (FIT)** is the withholding for federal income tax, based on the number of allowances you chose on Form W-4.

- ✔ **Social Security** is the retirement component of Social Security, designed to pay you a monthly income later in your life.

- ✔ **Medicare** is the Medicare component of Social Security, designed to provide payments for healthcare later in your life.

- ✔ **State income tax (SIT)** is the withholding for state income tax, based on the number of allowances you chose on Form W-4.

- ✔ **State disability insurance (SDI)** is a premium you pay, in case you're injured on the job.

Other deductions are usually voluntary and might include deductions for a retirement plan, such as a 401(k), SIMPLE IRA, or SEP IRA; health savings account; or voluntary contributions to a charity.

Deduction math is simple: Simply subtract your deductions from your gross pay. What's left is your net pay.

net pay = gross pay – deductions

Employee				SSN
George Spelvin	(President), 552 Brock Road, Nevada City, CA 95959			***_**
				Pay P

Earnings and Hours	Qty	Rate	Current	YTD Amount
Hourly Regular Rate	32:00	41.00	1,312.00	17,056.00

Deductions From Gross			Current	YTD Amount
SIMPLE IRA Employee			−196.80	−2,558.40

Taxes			Current	YTD Amount
FIT Federal Withholding			−168.00	−2,312.00
Social Security Employee			−81.34	−1,057.42
Medicare Employee			−19.02	−247.26
CA – SIT Withholding			−49.47	−643.11
CA – SDI Disability Employee			−14.43	−171.87
			−332.26	−4,431.66

Net Pay			782.94	10,065.94

Non-taxable Company Items			Current	YTD Amount
SIMPLE IRA Company			39.36	511.68

Illustration by Wiley, Composition Services Graphics

Figure 13-3: Typical payroll deductions.

Using Form W-4 to change how much is withheld

IRS Form W-4 (official name: Employee's Withholding Allowance Certificate) is a simple form that tells your employer how much money to withhold for federal income tax (see Figure 13-4).

Figure 13-4: IRS Form W-4.

You don't tell your company how much money to withhold. (It has tables and computers to do that task.) Instead, you use a Form W-4 to tell you're your company how many *allowances* to select. The higher the number of allowances, the less federal income tax your employer withholds. Less withheld means you have more money to spend.

The number you enter is usually based on the total number of dependents in your household. Generally, you declare yourself, your spouse, and your children. There are variations and other tweaks as well. If your mother lives with you, for example, you can usually claim her, too. If you declare yourself, your spouse, your two minor kids, and your mom who lives with you, you have 5 allowances.

You complete Form W-4 when you're hired for a job. But you can fill out a new one any time. If you have a new baby, for example, you'd want to increase your number of allowances. If you think you'll take a hit at tax time, you can decrease your number of allowances so that more taxes are withheld each paycheck.

The objective is generally to free as much cash as possible *now*. There's usually no point in over-withholding. If you'd like to do a withholding calculation using the W-4 worksheet, you can download a W-4 form from http://www.irs.gov.

Chapter 14

How Taxing! (Almost) Understanding the Government

- -

In This Chapter

▶ Comprehending income tax

▶ Taking on the terrifying Form 1040

▶ Managing the many kinds of taxes

▶ Reviewing government fees

- -

*I*t takes money to run the government, and money comes to it in the form of taxes. A *tax* is a financial charge that the government imposes on a taxpayer, who may be an individual (you) or a corporation (your company).

The great Supreme Court Justice Oliver Wendell Holmes said, "Taxes are the price we pay for a civilized society." That makes sense, because government takes care of national defense, police services, firefighting, scientific research, and much, much more. At their best, taxes you pay return value in services to you. At their worst, taxes are a little like what Dr. Waldman says in the 1931 movie, *Frankenstein*: "You have created a monster, and it will destroy you!"

In this chapter, you see some top-level information about taxes — what they are, how they work, and how to calculate them.

Illuminating Income Taxes

An *income tax* is a tax on personal or corporate income. Chances are, you pay United States federal income tax, and many states (41 out of 50 at last count) levy income taxes, too.

Fortunately, income tax is easier to understand and calculate than many people think. You need only to know some terms and do simple arithmetic.

Emperor Wang Mang of China imposed an income tax in 10 AD, and Henry II used an income tax to raise money for the Third Crusade. These days, about 123 countries have a tax on income.

Taming Form 1040

The basic income tax form in the United States is the Internal Revenue Service Form 1040, U.S. Individual Income Tax Return (see Figure 14-1). At the core, the 1040 is simple. You figure out your total income, make a few subtractions, and pay a tax on the difference.

Figure 14-1: The top portion of Form 1040.

This section is about filling out a tax form. I'm not an accountant or an attorney, so for tax advice, you want to consult a tax professional.

Form 1040 has several sections, each of which contains a group of lines that you fill in or ignore, depending on what applies to you. In the next sections, I take you through the form, explaining the math along the way.

After putting in your name, address, other identifying information and indicating your filing status, you proceed through a series of sections.

The following sections are based on the 2011 1040 Form. So stay alert! In addition to the form changing every year, the standard deduction amount, exemption amount, and tax rates change every year, too. To avoid trouble, make sure you use the current form and the correct amounts.

The Exemptions section

Exemptions lower your taxable income (and therefore lower your tax). In the Exemptions section, you claim yourself, your spouse (if applicable), and your dependents if you have them, and add up all the people you claimed. For example, if you check boxes for yourself and your spouse, and list the names of your three kids, you have 5 exemptions.

The Income section

In this section, you tally up all the ways you brought in money to get your *total income*. Income has many sources, including your wages, interest, dividends, and alimony. This is also where self-employment income and farm income go. Even unemployment compensation (if you can believe it!) is subject to income tax. But don't worry (yet). By contrast, *taxable income* is income you will pay taxes on, which is going to be lower than total income. Most people get their income figure from Form W-2, which their company gives them at the end of the year.

The Adjusted Gross Income section

Adjusted gross income (AGI) is just what it says: income adjusted by adjustments. In this section, you record any adjustments you're entitled to. *Adjustments* reduce your AGI (which will reduce your taxable income). They include educator expenses (for teachers, of course), moving expenses, interest on student loans, and IRA contributions. You may have no adjustments, but if you do, add them up.

To calculate adjusted gross income, you do a simple subtraction:

adjusted gross income = income − adjustments

The Taxes and Credits section

This is where you enter deductions and exemptions and calculate your tax.

A *deduction* is an expense you're allowed to use to reduce AGI. That's good, because if you lower your AGI, the result will be a lower *taxable income*. There are many kinds of deductions, and the 1040 allows you to choose between a standard deduction (in 2011, $5,800 for a single person) or a larger amount, if you "itemize" deductions on Form 1040 Schedule A. An *exemption* isn't a deduction, but it has the same function — lowering your AGI.

Here are some important parts of the Taxes and Credits sections:

✔ **Entering your deductions:** Enter the standard deduction or the itemized deductions from Schedule A.

✔ **Tallying up your exemptions:** For every exemption you entered on page 1 (yourself, spouse, and dependents), you multiply by $3,700 (in 2011).

✔ **Finding your taxable income:** *Taxable income* is the income you'll pay taxes on. To figure it, you subtract your deductions and your exemptions from your AGI:

taxable income = AGI − itemized deductions − exemption amount

✔ **Indicating your tax and tax credits:** *Tax* is the tax you owe. You can get the number by doing a tax rate calculation, but most people simply go to the tax tables in the Form 1040 instructions. You enter the tax on the Tax line (which on 2011 forms was Line 44). Below that line are about eight other "tax credit" lines, where you might enter (for example) a residential energy credit.

You can be sure that this is a simple example. The 1040 has many other tax and credit lines. Just add tax credits together and subtract them from your tax.

No negative numbers exist in Government Land! If your tax is $1,000 and your credits are $2,000, *you enter 0!*

Marginal rate

The *marginal tax rate* is the rate on the last dollar of income you earned. Despite the murky name and equally murky definition, it's not a hard number to understand. The government taxes the first dollars you make in a year at a low rate. Then when you earn a few more, those additional dollars get taxed at a higher rate. By the end of the year, you may have earned enough for the last dollars to be taxed at an even higher rate.

The marginal rates are known as *tax brackets*. You may have heard a friend say, "I'm in the 28 percent tax bracket." In the U.S., the brackets are 10, 15, 25, 28, 33, and 35 percent. To see the current tax brackets and the income each applies to, visit http://www.irs.gov.

To determine your marginal rate, you can use the tax tables. Or use a marginal rate calculator. You can find both simple and complex calculators on the Internet. You enter wages, filing status, number of dependents, and the amount of itemized deductions. The calculator tells you your marginal tax rate. To see one of them, visit http://www.dinkytown.net/java/TaxMargin.html.

The Other Taxes section

There's always room for more taxes. Among other things, this section asks you to enter self-employment task (if you're in business for yourself). Just add up the lines and add the total to your tax. You put this in a line called "Total Tax." At last!

The Payments section

The amount of your federal withholding (from the Form W-2 your company gave you) goes in this section. There are a couple of other possible credits (such as a first-time homebuyer credit), too.

The Refund and Amount You Owe sections

If your withholding is larger than your total tax, you are going to get a refund. If it's smaller, you owe Uncle Sam some money. Just subtract.

Deciding whether to itemize

Should you itemize deductions? For many taxpayers, it's an annual question. The short answer is, if you own a home with a mortgage, probably yes. If you don't, you'll have to dig deeper to determine whether itemizing is the way to go. Here's why: If you don't itemize, you're entitled to a standard deduction. In 2011, that deduction was $5,800 for a single person and $11,600 for a married couple filing jointly.

The main reason to itemize is when you can get more than that standard deduction because you have more than $5,800 or $11,600 worth of deductions. The reason so many homeowners opt to itemize is because interest paid on a mortgage ends up being a big deduction that either pushes them over the standard deduction amount or gets them close enough to it that, with a few other deductions, they end up exceeding the standard deduction.

When you're making the "itemize or not" decision, your math task is to determine whether your deductions exceed the standard deduction amount. It's simple addition and comparison: Add up all your individual deductions, compare that amount to the standard deduction amount, and go with the one that's higher.

What's not so easy is determining which and how much of the potential individual deductions you can take. To itemize, you use IRS Form 1040 Schedule A (Figure 14-2 shows the top part of this form).

SCHEDULE A (Form 1040)	Itemized Deductions		OMB No. 1545-0074 2011
Department of the Treasury Internal Revenue Service (99)	▶ Attach to Form 1040.　▶ See Instructions for Schedule A (Form 1040).		Attachment Sequence No. 07
Name(s) shown on Form 1040			Your social security number
Medical and Dental Expenses	**Caution.** Do not include expenses reimbursed or paid by others.		
	1 Medical and dental expenses (see instructions)	1	
	2 Enter amount from Form 1040, line 38 ⎣ 2 ⎤		
	3 Multiply line 2 by 7.5% (.075)	3	
	4 Subtract line 3 from line 1. If line 3 is more than line 1, enter -0-		4
Taxes You Paid	5 State and local (check only **one** box):		
	a ☐ Income taxes, or ⎫	5	
	b ☐ General sales taxes ⎭		

Figure 14-2: IRS Form 1040 Schedule A.

The Schedule A deductions fall into several categories. The following list summarizes each category and what it takes to meet the deduction floor, or threshold.

A *floor* (also called a *threshold*) is a level below which no deduction is allowed. Different deductions have different floors. For example, job expenses and miscellaneous deductions are deductible only *to the extent that they exceed* 2 percent of your AGI. Below that, you get nada. For example, if you have an AGI of $20,000, 2 percent is $400 (simply multiply your AGI by 0.02). You may have $600 in job expenses, but you don't get $600 as a deduction. You get the amount over $400, or $200.

✔ **Medical and Dental Expenses:** You can claim medical and dental expenses, but here's the catch: You can deduct *only* the amount by which your total medical care expenses for the year exceed 7.5 percent of your AGI. (Don'tcha just love it? Some of those medical expenses you thought were deductible probably aren't, due to the threshold.) To determine how much expenses exceed the threshold, multiply your AGI by 0.075 to determine the floor. You can deduct any medical expenses over that amount.

The math you use here is entirely about calculating the floor and subtracting it to get your allowable medical expense deduction.

✔ **Taxes You Paid:** Here's the place to enter a variety of taxes, such as state income taxes, sales taxes, real estate taxes, and personal property taxes. Check with your tax advisor. Then add them all up.

✔ **Interest You Paid:** Homeowners get excited about this section. That's because most homeowners have a big interest component in their mortgage payments. The lender will send you a Form 1098, which gives you the amount of interest you paid. The home mortgage interest deduction is America's favorite tax deduction. It's usually big, and the lender does the math for you.

If you rent, you pay the rent, but the landlord takes the mortgage interest deduction.

✔ **Gifts to Charity:** Most charitable donations are small amounts of cash or are represented by a receipt for the items you donate to a charity thrift store.

If you're giving vast amounts of stock or property to charity, you definitely need to talk to a tax professional because how much charitable giving you can deduct is subject to many rules. There are often limitations regarding what you can give.

✔ **Casualty and Theft Losses:** Generally, you can deduct losses to your home, household goods, and motor vehicles. But, of course, limitations exist if you have insurance — and most people have automobile, homeowner's, or renter's insurance, if they can afford it. To figure it all out, you need an instruction pamphlet from the IRS, and you need to attach a special form (Form 4684) to your tax return. You can see the form at http://www.irs.gov/pub/irs-pdf/f4684.pdf.

✔ **Job Expenses and Certain Miscellaneous Deductions:** Some employee job expenses — travel, union dues, and job education — are deductible. You need to attach Form 2106 or

Form 2106-EZ. Again, your math is simple. Just list the items and add them up.

By the way, tax preparer fees are deductible, too. So's a safe deposit box and other such items.

✔ **Other Miscellaneous Deductions:** "Other" deductions can be an adventure. For example, a horse is likely to be disallowed as a "hobby expense," unless you're in the horse business. It doesn't matter how much Flicka, Black Beauty, or Stripes costs you in maintenance and training. Bottom line: Do your research and ask your tax professional. Then list the allowable items and add them up.

As they say in Hollywood, "It's a wrap!" Total the deductions from the various sections. The real-life math is just to add everything up and move the number over to the correct line on Form 1040.

Observing Other Taxes

As Ben Franklin noted, nothing is certain but death and taxes. And taxes of all kinds have been around for a long time. Taxes come in many forms, and you can easily calculate them if you know how.

Surveying sales tax

A *sales tax* is a percentage-based tax charged on the goods and services you buy. Almost all purchases (except some Internet sales) require that you pay sales tax. Sales taxes are state taxes, sometimes with a local tax added on.

In the United States, sales tax rates vary from 0 percent (New Hampshire, Delaware, Oregon, Montana, and Alaska) to 7.50 percent (California).

Although you can see how much you've paid in sales tax just by checking the tax line on your grocery receipt or restaurant check, occasionally you want to know how much tax you'll owe *before* you make a purchase. In that case, simply take the cost of the item and multiply it by the sales tax percentage. If you're buying a $139 dress and sales tax is 7 percent, use this equation: $139 \times 0.07 = \$9.73$. See Chapter 5 for the details on sales tax.

Processing property tax

Property tax is a local tax that homeowners pay and that is assessed by the county assessor. (Renters pay property tax, too, because landlords pass it on to them in the monthly rent.)

The actual amount you pay for property tax depends on your local tax rate (in California where I live it's 1.25 percent) and the assessed value of your home. For a $200,000 home in California, for example, you calculate taxes as follows:

tax = assessed value × 1.25 percent

tax = $200,000.00 × .0125

tax = $2,500.00

Many mortgage companies include property taxes in your monthly mortgage payment and keep the amount in escrow until a tax payment becomes due. If you own your home outright, however, the tax bill comes directly to you. Also keep in mind that property taxes are often levied twice a year. To determine what you'll have pony up each six months, simply divide your yearly tax burden by 2. If your annual property tax burden is $2,500, for example, you will pay $1,250 every six months. If you have an escrow account, about $208 of your monthly payment goes into it ($2,500 ÷ 12).

Fee, fie, foe, fum

Whereas a *tax* is for the general fund and has no direct connection with individual benefits, a government *fee* is money you pay for specific, directly beneficial goods or services. For example, a *user fee*, such as the one you pay to get into a national park, benefits you directly. If you don't want to go to the national park, you don't have to pay the fee.

Don't be misled by someone who calls everything, including fees, "taxes" — for example, insisting that bridge and tunnel tolls are taxes. They aren't. The reverse is sometimes true: Every once in a while, a state or local government tries to call a tax a fee. That's understandable, because a lot of people are tax-averse, but it's wrong, and the courts usually call them on it.

Following are some common fees:

- ✔ **Motor vehicle fees:** If you own a vehicle, you very likely pay registration fees every year to the Department of Motor Vehicles. In California (where I live and where we worship cars), a typical vehicle renewal notice bills out several fees, like registration fees, license fees (which you may get an income tax deduction for), fees for personalized license plates (called *vanity plates*), and so on.

- ✔ **Licenses and permit fees:** Sometimes a fee is part of a license or permit application. You can pay fees for building permits, demolition permits, logging permits, entry to the landfill, business licenses, and so forth. Fishing and hunting license fees fall into this category, too.

The math involved in figuring out what fees you owe is pretty simple: Find out what the current fee is, add in any subcomponents of the fee, and then add them up. For example, a California fishing license not only has the basic sport fishing component, but additional subcomponents for fishing in the ocean, using a second rod, fishing in the Colorado River, taking abalone, taking salmon in the northern rivers, taking steelhead, and taking spiny lobsters.

Making sense of government indexes

An *index* is a set of numbers, usually created by statisticians and evaluated by economists.

The U.S. government has many indexes, and they are mainly money-related numbers. In government indexes, math and civics merge: These numbers are bandied about and used to justify all sorts of policy decisions and to characterize how well or poorly the U.S. is doing. By understanding what these terms mean and how the numbers are calculated, you can better evaluate the info you're getting.

- ✔ **Gross National Product (GNP):** The GNP is the market value of all the "output" (products and services) produced in a year. It considers all the property and labor of the country's residents, but it doesn't care whether the enterprises are in the U.S. or abroad. So, if Apple makes iPads in China or HP makes PCs in Indonesia, that's part of the GNP. GNP was a hot number until 1991. Now GDP is used more frequently.

- ✔ **Gross Domestic Product (GDP):** GDP is the market value of all goods and services produced *within a country* in a period. A big growth rate is better than a small one, or a negative one. For example, India is doing fine with a GDP growth rate of 7.8 percent; the United States is recovering from a recession and has a growth rate of 1.7 percent; and Portugal is in big trouble with a growth rate of –2.2 percent.

By the way, the United States has the biggest economy in the world, with a GDP of $14.447 trillion. China is #2, with a GDP of $5.739 trillion.

✔ **GDP per capita:** This is a measure of a country's standard of living. It's a country's GDP divided by the country's population. You can find lists of GDP per capita on the Internet. Take a look. You might be surprised that Luxembourg gets the top spot, with a GDP per capita of $113,533 USD. Where's the good old U.S. of A? A bit down the list at number 14, with $48,387 USD per person.

✔ **Per capita income:** *Per capita income* (or *income per person*) is the average income of a country, a state, or a city. "Per capita" means "per head." To get per capita income, you take all sources of income and divide it by the population. Per capita income is a very rough measure of prosperity, and it doesn't show distribution. Therefore, it doesn't say anything about individual people, rich or poor.

✔ **Inflation rate:** The *inflation rate* is the annual rate of increase in the consumer price index (CPI). If prices go up, say, 3 percent in a year, the inflation rate is 3 percent. An example of inflation at work is when, over time, a loaf of bread increases in price from $2.00 to $4.00. Your bucks can't buy all the bread they used to, so you might also say that your *purchasing power* (the amount of goods or services you can buy with a unit of currency) has gone down. Of course, if your income goes up as prices go up, your purchasing power isn't diminished. You don't feel the pain (as much). That's "keeping up with inflation."

A famous American myth claims that a young George Washington threw a silver dollar across the Potomac River and inspired a very old joke from a 1942 Loony Tunes cartoon: "Why couldn't he do it today? Because a dollar doesn't go as far as it used to."

✔ **Unemployment rate:** The *unemployment rate* measures unemployment and is the percentage of the workforce that's not working, willing and able to work, and actively looking for work. In the United States, the unemployment rate doesn't reflect those working part time (the *underemployed*) and those who have given up looking for work after months or years of being unable to find a job.

The unemployment rate is calculated by dividing the number of unemployed workers by the total number of workers in the workforce and then multiplying that value by 100. For example, if a country with 300 million people has a workforce of 150 million people and 12 million people are unemployed, the unemployment rate is 8 percent ($12 \div 150 \times 100$).

✔ **Consumer price index:** The U.S. Consumer Price Index (CPI) shows changes in the price level of goods and services you buy. It's the main measure of inflation. Every month, the Bureau of Labor Statistics determines the CPI by looking at consumer goods and services, like housing, clothing, energy, and utilities. A low monthly CPI change (an increase of 0.3 percent, for example) is a good thing, as it means that prices are not rising fast.

Part IV
The Part of Tens

"Okay – let's play the statistical probabilities of this situation. There are 4 of us and 1 of him. Phillip will probably start screaming, Nora will probably faint, you'll probably yell at me for leaving the truck open, and there's a good probability I'll run like a weenie if he comes toward us."

In this part . . .

*T*he world loves lists of ten things, and in these chapters you find a bunch of fun and useful information in
a very small space. Here I list quick calculations you can
do in your head and fun games and activities that build or
use your math skills and sharpen your critical thinking.

Chapter 15

Ten Quick Calculations You Can Do in Your Head

In This Chapter

▶ Calculations that come in handy on the road

▶ Estimating taxes and tips on the fly

▶ Easy ways to determine pizza, paint, and other amounts

---

*W*hen a real-life math problem comes up, you may not have a calculator handy or the situation doesn't lend itself to using one (like when you're driving or when you're figuring the tip at a business lunch). Even if you do have one handy, the easiest and most efficient way to solve many math problems is in your head. Although not all math problems lend themselves to mental math, there are some handy calculations you can keep in your head. Here are ten of them.

Miles to Kilometers

To convert from kilometers to miles, you first need to know that 1 km = 0.62 mi (actually, it's really 0.621371, but don't worry about the other decimal places). Say you leave Amsterdam to visit Haarlem, a trip of 20 km. If 1 kilometer equals about 0.62 miles, then 10 km = 6.2 mi. Double that, and you'll see that the 20 km trip is 12.4 mi.

To go from miles to kilometers, divide the distance in miles by 2 and then add that number back to the original distance. For example, say your destination is 28 miles away. Divide 28 by 2 to get 14, and then add 14 back to the original number of miles (28). The answer: 42. (If you're interested, 1 mi = 1.60934. In this method, you're multiplying by 1.5, not the actual conversion factor of 1.6, but it's close enough.)

Miles to Your Destination

When you're driving, you can find the distance to your destination in three ways: You can look at the distances on the highway signs. You can enter your destination into your GPS and check it periodically to see how far you have left to go. Or you can do a little mental math: Before traveling, look up how far it is to your destination (you can use an online map program, like Google Maps — https://maps.google.com) and set your trip odometer to 0. As you drive, subtract the miles on the trip odometer (how far you've traveled) from the total distance you know you will travel.

Time to Destination

When you're driving, the easy answer to the question "Are we there yet?" is "No." A harder question to answer is, "When will we get there?" To find the time to your destination, you just need to know distance and speed, because time = distance ÷ speed.

For example, suppose that the road sign says your destination is 180 miles away, and your speedometer says you're moving along at about 60 miles per hour. Divide distance (180 miles) by speed (60 miles per hour) to get time (3 hours). Of course, for other distances and speeds, the calculations are harder, but the formula is the same.

Sales Tax and VAT

In many parts of the United States, sales tax on purchases is over 8 percent. To be safe, estimate it at 10 percent. To find out how much tax you'll pay on an item, just divide the item's price by 10 (move the decimal over 1 place to the left). For a $15.00 item, for example, the result is $1.50.

Many countries have a value added tax (VAT), and you use the same kind of math to figure those. The VAT is usually about 20 percent of the price of an item. To estimate the VAT, first divide the price by 10 (for a €50.00 item, for example, that's €5.00) and then double it. The VAT on the €50.00 item will be €10.00.

Tips

To leave a 15 percent tip, you simply figure out what 10 percent of the check is (move the decimal to the left 1 place), what 5 percent of the check is (halve the 10 percent value), and then add the numbers together. For example, a $23.00 check divided by 10 is $2.30. Divide that amount by 2 to get$ 1.15. Now add the two numbers together: $2.30 + $1.15 = $3.45.

To leave a 20 percent tip, move the decimal place over 1 spot to determine what 10 percent of the bill is and then double that amount. For example, 10 percent of a $55.00 check is $5.50. Double that to get $11.00. Easy peasy.

How Much Paint to Buy

To figure out how much paint to buy, first divide the area a gallon will cover (you can get this information off the paint can) by the height of a room. This gives you the number of linear feet the gallon will cover. If the paint covers 400 square feet and your walls are 8 feet high, you know that a gallon is enough to paint 50 linear feet. Next, measure the perimeter of the room. For a 9 foot by 12 foot room, for example, the perimeter is 9 + 9 + 12 + 12 feet, or 42 feet. If a gallon covers 50 linear feet, you know that one gallon will do the job with some paint left over. You can extend this to multiple rooms. Just add all the perimeters together and count on using 1 gallon of paint for every 50 linear feet.

Number of Pizzas to Buy

Pizza is a favorite food when you're feeding a bunch of people, whether it's your kid's Little League team or a project team that's working late on a deadline. The math comes in when you have to determine how many pizzas to buy. Simply figure out how many pieces total your group is likely to eat and then divide that number by the number of slices in a pizza.

Here's an example: Say you're taking 12 kids from your youth soccer team out for pizza. You figure that each kid will eat 2 pieces, so you know you need 24 pieces of pizza ($12 \times 2 = 24$), and a typical 14-inch pizza has 8 slices. By dividing the total number of slices needed by the number of slices in a pizza, you know you need 3 pizzas ($24 \div 8 = 3$). You can use the same kind of calculation to figure out how much cake to buy for a wedding shower, how many boxes of juice to buy for your kid's class party, and other similar scenarios.

Blood Alcohol Content (BAC)

In most states, driving with a blood alcohol content (BAC) — the percentage of alcohol in your blood — of 0.08 percent or higher will get you in serious trouble. So how much can you consume before you cross the legal drinking threshold?

A 12 ounce can of beer, a 5 ounce glass of wine, and a 1 ounce shot of alcohol all have about the same impact on your BAC. They each add about 0.02 percent to your BAC. The smart math here is to go to your state's Department of Motor Vehicles website and find a BAC chart. These charts usually show gender, weight, and the number of drinks that will mess you up. Memorize the numbers that apply to you, and should you drink, count your drinks and then stop. Also, you can find a very good online BAC calculator at `http://bloodalcoholcalculator.org`.

Never operate a motor vehicle (including a boat) if you're impaired by alcohol. The idea is to drink very moderately, if at all. You'll be a safer driver, and you'll avoid an arrest for driving under the influence (DUI) or driving while impaired (DWI).

Dollars to Pounds or Euros

The *exchange rate* between currencies changes a little every day, but in general, a pound (£ or GBP) is worth about $1.50 and a euro (€ or EUR) is worth about $1.20. To convert from euros to dollars, multiply any price by 1.2. For example, if you see an item priced at €10.00, that's the equivalent of $12.00 ($10.00 \times 1.2 = 12.00$). A €25.00 item works out to be $30.00. To convert from pounds to dollars, multiply the price by 1.5. For example, if you see an item priced at £10.00, that's the equivalent of $15.00 ($10.00 \times 1.5 = 15.00$). A £25.00 item is $37.50.

TIP

If you carry a smartphone that works in Europe, use a currency converter app to find the exact exchange rate. Or you can just ask the staff at your hotel.

Gas Mileage

You can estimate gas mileage two ways: The first way is to divide the miles you can drive on a tank of fuel by the capacity of the tank. But this method is imprecise for a variety of reasons, two being that you never really know what a "full" tank is and it doesn't account for driving conditions.

A quick and easy way to calculate miles per gallon is to do a little test run. Say you have a 12-gallon gas tank (info you can get from your owner's manual). Fill the tank up, set your trip odometer to zero, and take a short day trip. When the fuel gauge registers 3/4, look at the trip meter to see how far you've gone. If the gauge says 90 miles, you have driven 90 miles on 1/4 tank of gas, or about three gallons. Divide 90 by 3 to find that you're getting about 30 mpg.

Chapter 16

Ten Activities That Build Math Skills

● ●

In This Chapter

▶ Performing powerful parlor tricks

▶ Doing what is logical, Captain

▶ Solving math puzzles and playing math games online

● ●

Man (and woman) doth not live by bread alone. That is, life involves more than work and chores. So what do you do with your leisure time? When there's nothing on TV, try reading and games. Reading entertains, educates, and edifies. However, math games help to make you a math whiz. Here are ten games and math-related activities that can get you there.

Playing Sudoku

Sudoku is a number placement puzzle that requires logic and a quick grasp of the numbers 1–9. There's no adding, subtracting, or any other math operation. You simply have to determine what number is and isn't in a row, column, or subregion. It's as simple as tic-tac-toe, but it's an excellent challenge for the thinking person.

The easy Sudoku games are fast; the hard ones, not so much. You can find Sudoku puzzles in the newspaper, online, in apps for smartphone or tablet, and even in in-flight magazines on the airline.

Playing Elementary Math Games on the Internet

You could go to many Internet sites to learn math rules, formulas, and step-by-step procedures. But life short, so instead play some math games. Visit http://www.coolmath-games.com.

Here you'll find money games, number games, logic games, and much more. All free! And if you think these "children's" games are easy to master, try playing a few. Even "Lemonade Stand" can be a difficult business resource management exercise.

Working through Logic Puzzles

You may have grown up with logic puzzles (you know, the kind that asks you to match up people with houses and pets). The typical result is something like "The Swede lives in the red house and owns a dog." At first, these games strain your brain, as they should. Later, your skills reduce the pain and you get a gain.

Try Sherlock, one of the best of the logic puzzles. Sherlock and many other logic games are at Everett Kaser's website (http://www.kaser.com). In my opinion, he's the undisputed master of logic puzzles.

Noting the Birthday Paradox

The Birthday Paradox isn't so much a game as a statistical probability that will baffle and amaze your friends. Basically, this paradox states that, in a room full of people, there's an excellent chance that two of them have the same birthday. When you have 367 people in the room, the probability is 100 percent, because there are at the most 366 days in a year. Interestingly, probability calculations show that the likelihood of a match is 99 percent when only 57 people are in the room. And you might win a bet or two knowing this: The probability is 50 percent when only 23 people in the room.

That 50 percent probability is as good as a coin flip, but the birthday paradox is much more dramatic!

Knowing the Value of Pi

Knowing the mathematical constant pi (π) may not have a lot of practical uses unless you're a mathematician or a scientist, but it puts you in a relatively small group of people who, upon hearing the word "pi," don't think immediately of apple or pecan.

Here are three things that any self-respecting pi lover needs to know: First, pi allows you to the calculate areas and circumferences of circles and the volumes of cylinders and spheres. Second, pi's value is approximately 3.14, but no one knows the absolute value of pi because the decimal places go on forever. (If you want extra credit, memorize pi to 12 decimal places: 3.141592653589.) Third, March 14th (3/14) of every year is National Pi Day.

Guessing a Friend's Age

Using a technique called a binary search, which comes from computer science, you can guess a friend's age in a maximum of seven tries. To start, tell a friend, "I can guess your age in seven guesses or less. I'll name a number, and you say 'high,' 'low,' or 'yes.'"

Start in the middle of a range of age values. If you were using the ages 1–64, you'd start with 32 and then continue to halve the search range at every "high" or "low" response.

Say for example that, when you say "32," your friend says "Low." You halve the search range (which now becomes 33–64) and pick the number in the middle of the new range: "48." If the answer isn't "yes," halve the search range again. Worst case, it'll take you seven guesses to get to the right age.

You can guess the age in even fewer tries if the person is obviously not a child or teenager. Start the search range at age 20.

Playing Hidden Object Games

A hidden object game (HOG) is a type of computer game where you have to find items from a list that are hidden in a scene, and there are numerous scenes. Finding the objects advances a story, and in the story, you need to place found objects to make something happen. The big game is interspersed with fun mini-games, too.

If a hidden object game doesn't test your investigative, reasoning, and sequencing skills, it's hard to say what will.

The games are very inexpensive, and you can extend play over hours or days. You can find these games in lots of places, but one of the best is Big Fish Games. Visit http://www.bigfishgames.com.

Flipping Coins

Flipping a coin consist of tossing a coin into the air so it rotates several times. In some games, a participant "calls" it in the air as coming up "heads" or "tails." Play a simple (almost mindless) game to learn a lesson in probability.

Flipping a coin has two outcomes — heads or tails. The probability of one side coming up is 1 in 2, or 50 percent. Flip a coin a few dozen times, and you'll see that each side comes up about half the time. So what happens if the coin comes up heads 100 times in a row (for example)? That must mean that the next flip has to be tails. Nope, the various flips are each independent events, so the probability is 50 percent *every time*.

The Romans called coin flipping *navia aut caput* ("ship or head"), because that's what was on their coins. They considered the outcome to be expression of divine will. Today, a coin flip ("the toss") is used in football games to decide which team gets first use of the ball.

Playing Games with Your Kids

Even very young children benefit from math games, and when you help them, you improve your skills, too. Begin with simple sequencing games, such "This Little Piggy" ("This little piggy went to market; this little piggy stayed home . . ." and so forth). Later, you and your child can sing "This Old Man," ("This old man, he played one, he played knick-knack on my thumb") together. After that, use Sesame Street counting games to help teach both math skills and computer skills. Visit http://pbskids.org/games/counting.html.

Playing Angry Birds

Angry Birds is one of the most popular games for smartphones and tablets ever made. Over 12,000,000 copies have been purchased at the Apple App Store, and it's available for Android devices, too.

Angry Birds is excellent for developing your mental math skills, and you won't even know you're doing so. You figure trajectories without thinking very much about them. Economy is important, too, as you get more points for knocking down the pigs' structures, using as few birds as possible.

Who knew that shooting wingless birds at pigs using a slingshot would excite so many people? And the "purchase" price is $0.00, unless you upgrade.

Index

• *G* •

Apple & Mac

iPad 2 For Dummies,
3rd Edition
978-1-118-17679-5

iPhone 4S
For Dummies,
5th Edition
978-1-118-03671-6

iPod touch For
Dummies, 3rd Edition
978-1-118-12960-9

Mac OS X Lion
For Dummies
978-1-118-02205-4

Blogging & Social Media

CityVille For Dummies
978-1-118-08337-6

Facebook For Dummies,
4th Edition
978-1-118-09562-1

Mom Blogging
For Dummies
978-1-118-03843-7

Twitter For Dummies,
2nd Edition
978-0-470-76879-2

WordPress For
Dummies, 4th Edition
978-1-118-07342-1

Business

Cash Flow For Dummies
978-1-118-01850-7

Investing For Dummies,
6th Edition
978-0-470-90545-6

Job Searching with
Social Media
For Dummies
978-0-470-93072-4

QuickBooks 2012
For Dummies
978-1-118-09120-3

Resumes
For Dummies,
6th Edition
978-0-470-87361-8

Starting an Etsy
Business For Dummies
978-0-470-93067-0

Cooking & Entertaining

Cooking Basics
For Dummies,
4th Edition
978-0-470-91388-8

Wine For Dummies,
4th Edition
978-0-470-04579-4

Diet & Nutrition

Kettlebells
For Dummies
978-0-470-59929-7

Nutrition For Dummies,
5th Edition
978-0-470-93231-5

Restaurant Calorie
Counter For Dummies,
2nd Edition
978-0-470-64405-8

Digital Photography

Digital SLR Cameras
& Photography For
Dummies, 4th Edition
978-1-118-14489-3

Digital SLR Settings
& Shortcuts
For Dummies
978-0-470-91763-3

Photoshop Elements 10
For Dummies
978-1-118-10742-3

Gardening

Gardening Basics
For Dummies
978-0-470-03749-2

Vegetable Gardening
For Dummies,
2nd Edition
978-0-470-49870-5

Green/Sustainable

Raising Chickens
For Dummies
978-0-470-46544-8

Green Cleaning
For Dummies
978-0-470-39106-8

Health

Diabetes For Dummies,
3rd Edition
978-0-470-27086-8

Food Allergies
For Dummies
978-0-470-09584-3

Living Gluten-Free
For Dummies,
2nd Edition
978-0-470-58589-4

Hobbies

Beekeeping
For Dummies,
2nd Edition
978-0-470-43065-1

Chess For Dummies,
3rd Edition
978-1-118-01695-4

Drawing For Dummies,
2nd Edition
978-0-470-61842-4

eBay For Dummies,
7th Edition
978-1-118-09806-6

Knitting For Dummies,
2nd Edition
978-0-470-28747-7

Language & Foreign Language

English Grammar
For Dummies,
2nd Edition
978-0-470-54664-2

French For Dummies,
2nd Edition
978-1-118-00464-7

German For Dummies,
2nd Edition
978-0-470-90101-4

Spanish Essentials
For Dummies
978-0-470-63751-7

Spanish For Dummies,
2nd Edition
978-0-470-87855-2

Math & Science

Algebra I For Dummies,
2nd Edition
978-0-470-55964-2

Biology For Dummies,
2nd Edition
978-0-470-59875-7

Chemistry For
Dummies, 2nd Edition
978-1-1180-0730-3

Geometry For Dummies,
2nd Edition
978-0-470-08946-0

Pre-Algebra Essentials
For Dummies
978-0-470-61838-7

Microsoft Office

Excel 2010 For
Dummies
978-0-470-48953-6

Office 2010 All-in-One
For Dummies
978-0-470-49748-7

Office 2011 for Mac
For Dummies
978-0-470-87869-9

Word 2010
For Dummies
978-0-470-48772-3

Music

Guitar For Dummies,
2nd Edition
978-0-7645-9904-0

Clarinet For Dummies
978-0-470-58477-4

iPod & iTunes
For Dummies,
9th Edition
978-1-118-13060-5

Pets

Cats For Dummies,
2nd Edition
978-0-7645-5275-5

Dogs All-in One
For Dummies
978-0470-52978-2

Saltwater Aquariums
For Dummies
978-0-470-06805-2

Religion & Inspiration

The Bible For Dummies
978-0-7645-5296-0

Catholicism For
Dummies, 2nd Edition
978-1-118-07778-8

Spirituality For
Dummies, 2nd Edition
978-0-470-19142-2

Self-Help & Relationships

Happiness For Dummies
978-0-470-28171-0

Overcoming Anxiety
For Dummies,
2nd Edition
978-0-470-57441-6

Seniors

Crosswords For Seniors
For Dummies
978-0-470-49157-7

iPad 2 For Seniors
For Dummies, 3rd
Edition
978-1-118-17678-8

Laptops & Tablets
For Seniors For
Dummies, 2nd Edition
978-1-118-09596-6

Smartphones & Tablets

BlackBerry For
Dummies, 5th Edition
978-1-118-10035-6

Droid X2 For Dummies
978-1-118-14864-8

HTC ThunderBolt
For Dummies
978-1-118-07601-9

MOTOROLA XOOM
For Dummies
978-1-118-08835-7

Sports

Basketball For
Dummies, 3rd Edition
978-1-118-07374-2

Football For Dummies,
2nd Edition
978-1-118-01261-1

Golf For Dummies,
4th Edition
978-0-470-88279-5

Test Prep

ACT For Dummies,
5th Edition
978-1-118-01259-8

ASVAB For Dummies,
3rd Edition
978-0-470-63760-9

The GRE Test For
Dummies, 7th Edition
978-0-470-00919-2

Police Officer Exam
For Dummies
978-0-470-88724-0

Series 7 Exam
For Dummies
978-0-470-09932-2

Web Development

HTML, CSS, & XHTML
For Dummies, 7th
Edition
978-0-470-91659-9

Drupal For Dummies,
2nd Edition
978-1-118-08348-2

Windows 7

Windows 7
For Dummies
978-0-470-49743-2

Windows 7
For Dummies,
Book + DVD Bundle
978-0-470-52398-8

Windows 7 All-in-One
For Dummies
978-0-470-48763-1